FURRY LOGIC

Also available in the Bloomsbury Sigma series:

FURRY LOGIC

THE PHYSICS OF ANIMAL LIFE

Matin Durrani

&

Liz Kalaugher

BLOOMSBURY
sigma

Bloomsbury Sigma
An imprint of Bloomsbury Publishing Plc

50 Bedford Square
London
WC1B 3DP
UK

1385 Broadway
New York
NY 10018
USA

www.bloomsbury.com

BLOOMSBURY and the Diana logo are trademarks of
Bloomsbury Publishing Plc

First published 2016

Photo credits (t = top, b = bottom, l = left, r = right, c = centre)
Colour section: P. 1: Daniel Mihailescu / Gettyimages (t); Micaela Pilotto / Pedro Reis / Roman
Stocker (bl); Sandra Standridge / Gettyimages (br). P. 2: Wolfgang Kaehler / Gettyimages (t);
Photo Researchers / FLPA (cr); Steven Trainoff Ph.D. / Gettyimages (b). P. 3: Michael Nitzschke /
Gettyimages (t); Keita Tanaka (cr); Hans Lang / Gettyimages (b). P. 4: Shin. T / Gettyimages (t);
Stephen Doggett, Department of Medical Entomology, Westmead Hospital, Sydney (tcr);
Tim Nowack Photography (bcr); Dirk Zabinsky / EyeEm / Gettyimages (b). P. 5: Alyssa Stark (t);
Tennessee Tech University (b). P. 6: Tracy Langkilde (t) A. & J. Visage / Gettyimages (b). P. 7: J. Rogers
Brothers (t, cr); Afriadi Hikmal / Gettyimages (b). P. 8: Thomas Ozanne / Gettyimages.

British Library Cataloguing-in-Publication Data
A catalogue record for this book is available from the British Library.

Library of Congress Cataloguing-in-Publication data has been applied for.

ISBN (hardback) 978-1-4729-1409-5
ISBN (ebook) 978-1-4729-1410-1

2 4 6 8 10 9 7 5 3 1

Chapter illustrations by Aaron Gregory
Diagrams by Marc Dando

Typeset in Bembo Std by Deanta Global Publishing Services, Chennai, India
Printed and bound in Great Britain by CPI Group (UK) Ltd, Croydon CR0 4YY

Bloomsbury Sigma, Book Nineteen

MIX
Paper from
responsible sources
FSC
www.fsc.org   FSC® C020471

To find out more about our authors and books visit www.bloomsbury.com.
Here you will find extracts, author interviews, details of forthcoming
events and the option to sign up for our newsletters.

To my parents, Saeed and Inge
And to Katja, Chiara and Alex   *Matin*

To Sue, Patrick, Mags, Catherine,
Justin, Daniel, Tom and Josh   *Liz*

Contents

INTRODUCTION
Furry Physics

It's tough being an animal. There's no central heating or
air-conditioning to keep you at a safe temperature, no
supermarket to provide supplies when you're peckish, and
no walls to protect you. Jump into a river to catch fish and
there won't be a towel ready to dry you when you wade
soggily out, your body chilling fast. To survive, animals must
use their senses, their wits, their mates, relations and pack-
members (except for solitary species like the leopard), as
well as their bodies, which have evolved over many years to
suit their living conditions. And that's where physics comes
in. Only recently have biologists and physicists realised just
how impressive animals are when it comes to exploiting
physics in their daily business of eating, drinking, mating and
generally avoiding being killed. Even wet pet dogs use

physics to shake themselves dry and soak anyone not quick enough to move away.

It's not that animals have worked out the principles of this science and designed their own bodies accordingly. It's that evolution has, over time, by gradual trial and error, come up with real-world systems that function well, using the science, principles and laws that humans call physics.

Animals got there first. The electric eel was firing off bursts of high voltage to kill crabs (see Chapter 5) and applying the principles of electricity long before scientists knew what it was. The eel doesn't understand electric currents, but we don't need to know anything about transistors or circuits to use a smartphone either. As long as the phone is smart, we don't have to be.

Worms without the wormholes

Before we kick off, a word of reassurance. This book is about how selected animals use physics to survive in the wild. If you're scared of physics, don't worry, we've kept things simple. You won't need to be an Einstein to follow what's going on. Don't expect anything weird like dark energy, Higgs bosons or wormholes – although we do mention worms at one point, or at the very least, snakes. On the other hand, if you are into physics, you'll be astounded by how often and how niftily your favourite science crops up in the animal world. From furry cats and dogs to spiny lobsters, mosquitoes and giant squid, physics is everywhere.

For physics lovers, the key thing to remember about biology is that pretty much everything is centred around sex or food. Physics, despite its obsession with the Big Bang, less so. For their species to survive, animals must pass on their genes by producing young. In almost all cases they need food to live long enough to breed, and perhaps to give them the energy to care for their young so that they too last into adulthood. A notable exception is the male fig wasp. This eats in its larval stage inside the fig, but once it metamorphoses into an adult it can no longer eat as its mouthparts are withered. Its sole purpose is to mate before it runs out of energy and dies.

If biology's your bag, the key thing to remember about physics is that it's much easier than biology. Honestly. If you're in a lab you have more control over your experiment. If you want to change just one thing (or one 'variable' in science-speak) to test its clout, it's a lot easier to do that inside a nice temperature- and humidity-controlled laboratory building sheltered from the weather than it is in a jungle. Or even a wildflower meadow, or, as we find out in Chapter 4, a zoo. And if you take the animal you're studying out of its environment, you don't know if that's made it behave differently. Leave that animal in its home, however, and you can't be sure if a variable you've changed has also altered something else – another animal or another variable – that you don't know about. And this might change your results without you being aware. So: biology hard, physics easy.

Now for the disclaimer: in *Furry Logic* sometimes we anthropomorphise, putting ourselves inside the heads of animals as if they were human. Biologists don't like this but it's easier to tell stories that way and so we won't apologise, or only a little anyway. And sometimes, whisper it, we simplify the physics a shade so that it doesn't block the narrative.

Some people turn to popular science seeking order and logic in today's messy world. But life's complicated. Sometimes the more you look at something the more complex it gets. It's like enjoying the beautiful colour and delicate scent of a rose, then examining it in close-up and seeing the velvety bloom on the petals, the veins at their pale base, the complex mini-forest of stamens and pollen dust at the centre, and the tiny hairs on the leaves underneath the bowl of the flower. Use a powerful microscope and you'll see the biological structures that make the rose work – its plumbing and individual cells. Unless you're seriously into botany, before you know it you've moved from pleasure and wonder to a bewildering world of fancy names and a flower that doesn't even look nice any more. It can be the same with physics explanations – there are levels upon levels, some that please everyone and others that are best left to the super-enthused and the genius. Here we've aimed for a level that remains fun; sometimes this means glossing over the geekiest details in the interests of beauty and simplicity, and we hope you're happy with that. If not, you'll probably enjoy looking up the equations and finer points yourself.

This book isn't an exhaustive account of the behaviour and characteristics of every single animal that applies physics. That would make it very long. Instead, each chapter covers a certain aspect of the science – heat, forces, fluids, sound, electricity and magnetism, light – and showcases its basic principles through the activities of a hand-picked set of animals. As the focus is on how animals use physics in their daily lives, we've chosen creatures, from peacocks to octopus and from elephants to bees, that actively take advantage of physics in drinking, catching their food, regulating their body temperature, defending themselves and more. It's more of a 'howdunit' than a whodunit, though you're unlikely to guess some of the answers.

One thing we don't look at, simply because there's enough material for a book in its own right, is how humans are exploiting their knowledge of the animal world for their own devices. So we don't cover how physicists developed heat sensors inspired by the structure of butterfly wings or created Velcro by studying how burrs stick to a dog's coat. Known as 'biomimetics' or 'bioinspiration', that field is interesting but well trodden. We do, however, become distracted – briefly – by a hearing aid based on elephant communication. Another topic out of our remit – again because others have tackled it – is the 'collective behaviour' of animals, such as flocks of birds flying in unison, the movement of penguins through a huddle, or ants working together to build a raft.

In our book the individual animals are the stars of the show.

CHAPTER ONE

Heat: The Warm-up Chapter

GENDER-SWAPPING SNAKES * FLOPPY-SKINNED DOGS
* MOSQUITOES THAT WEE BLOOD * KILLER BEES
* HOT-TAILED SQUIRRELS * VIPERS THAT 'SEE' HEAT
* BEETLES THAT 'HEAR' INFRARED

It's getting hot in here

In *Indiana Jones and the Raiders of the Lost Ark*, dashing archaeologist Henry 'Indiana' Jones, played by Harrison Ford, faces his worst nightmare. Terrified of snakes, he must brave a secret Egyptian chamber teeming with the reptiles to stop the Ark of the Covenant falling into enemy hands. As in many movies, the scene draws on this animal's classic image as a creature of both evil and power.

Steven Spielberg, however, had more than symbolism on his mind. After scouring every pet shop in London for snakes, the movie director's staff had to cut up rubber hoses to make up the numbers. Even some of the 'snakes' weren't snakes but legless lizards, a difference that's crucial to a biologist, if not to a desperate film crew. Like the slow worm in your compost heap, legless lizards are – as their name suggests – lizards whose legs have shrunk or disappeared.

The actors' motto 'never work with children or animals' could have been coined with snakes in mind. These reptiles bite. They slither. They're scary. But it's not just film-makers who have problems. Biologists studying snakes in the wild have a tricky time too. Snakes are hard to track down, and once a snake has spotted you it'll slide away or, worse, inject or spray you with venom that could kill if it gets under your skin or into your eyes.

Luckily for our story's non-phobic hero, Rick Shine from the University of Sydney in Australia, one snake is an exception to this 'difficult-to-work-with' rule. Catch it at the right moment and it barely cares if you pick it up. Shine could, if he wanted to, put these reptiles in a car and take them for a ride. Up to a point, as we'll find out later, he did. In autumn, winter and spring the red-sided garter snake (*Thamnophis sirtalis parietalis*) hangs out, like Indy's nemeses in *Raiders of the Lost Ark*, in huge groups, sometimes tens of thousands strong (there's a number to make a film director jealous). They won't be in a secret Egyptian chamber, but in limestone cracks under the frozen soil of the Canadian prairies in the province of Manitoba. For this snake is a record-breaker: it's the most northerly-living reptile in the western hemisphere.

Living where temperatures plummet to –40°C and snow coats the ground for eight or nine months a year seems crazy. Reptiles are ectotherms (from the Greek for 'heat from the outside') and can't generate their own body heat by burning food. Instead they rely on outside sources like the Sun, basking in its rays until they're warm enough to move fast and reproduce. Faced with freezing conditions, red-sided garter snakes huddle together for warmth in their winter hidey-holes and brumate, the snake equivalent of hibernation.

But being in Manitoba brings benefits for the red-sided reptiles, and for those studying them too. For a start, once it arrives, summer is warm, with temperatures touching 30°C. In April or May the snakes emerge and writhe around on the barren soil in groups hundreds or thousands strong. This sight, which looks like a giant tangle of squirming spaghetti, has intrigued people for years. What are the snakes up to?

With a plot that even Spielberg would be proud of, the mystery of the red-sided garter snake involves cool physics, lots of sex and a soupçon of gender-swapping. Not among Shine and his colleagues, we must stress, but the snakes themselves.

Great garters

Where are our manners? We should get to know this snake before we pry into its sex life. First let's meet the wider family. Garter snakes live throughout North America, although only those species that dwell where winters are extra-cold brumate. You'll find these reptiles in woods, forests or grasslands as long as there's water nearby. About half a metre long, they're venomous enough to kill small prey but not humans. Favourite snacks include frogs and fish, though the snakes will also feed on earthworms, rodents and small birds.

As for red-sided garter snakes, these reptiles don't, at first glance, live up to their name; they're black with cream stripes running the length of their body. Their red sides lie beneath overlapping scales and you can only see them if the snake puffs up its body in annoyance. During Manitoba's three or four months of summer, the reptiles make the most of the warmth and can stray more than 15km (9 miles) from their dens in search of food.

When the first chill hits the air – in August, no less – the snakes head back to their bunkers. At first they stay down there only at night or when it's cloudy. Once the daytime temperature drops below freezing, however, the reptiles put themselves under house arrest and snuggle together ready for nine months of cold. Their winter homes lie 6m (20ft) underground, below the frostline. At 10°C, the 'indoor' temperature is no summer's day but balmier than the -40°C outside. While they brumate, the snakes barely expend any energy, existing almost in suspended animation. They eat nothing and hardly breathe, getting up only now and then for a drink of water.

All of a slither

Stand by one of the snake pits in or around the village of Narcisse in late spring, the long-awaited sunshine warming your face, and you'll enjoy one of the most unusual sights in nature. Facing you will be a writhing carpet of mud-caked reptiles that have just emerged from their burrow and are huddling together once more. Stare closely and you'll notice something even more odd: almost all the snakes are males. At about 45cm (18in) long, they're some 15cm (6in) shorter than the females.

Undaunted by their smaller size, the male snakes venture outside several weeks before the females. By lying in wait, each hopes to be first to mate. As they slither past one another, the early-risers flick their tongues in search of chemicals called pheromones that the females release through their skin. After nine long months of brumation, sex seems to be the males' number one aim.

But there's a hitch. As soon as the females emerge from their lair, most leg it (as far as that's possible for a snake). Any who are slow off the mark become the centre of attention in a frenzied mating ball of tens or even hundreds of amorous males, each trying to loop his body around her so he's in the right position to mate. The female finds this stressful and does what she can to escape. With males outnumbering females by 10 or more to one, a male's chances of reproducing are slim.

The giant tangle of male snakes and these smaller mating balls are freaky enough, but something even weirder's going on. Look carefully and from time to time you'll see males give their full attention not to a female, but to another male. We're not being sexist but some males' behaviour is most ungentlemanly. Literally. He-snakes pretend to be she-snakes, or 'she-males' in the scientific lingo, giving off pheromones to impersonate females. She-males are easy to spot: they're the same length as other males but, having slithered from underground later, are still coated in mud. Rarely courting 'other' females, these transgender snakes crawl around sluggishly instead. Soon the 'real' males jump on them.

Harder than identifying the she-males is understanding what they're up to. If he wants to mate with a female, why does a male pretend to be the same sex as her? This puzzle set biologists scratching their heads. Perhaps becoming a she-male gives a male a reproductive edge so he can steal sperm from other males or avoid attack from larger rivals. But Rick Shine wondered if hanging out in a giant heap isn't only about reproduction. Could it also be a matter of heat?

Reptiles in the bag

Fortunately biology was on the researchers' side. You'd think desperate-to-mate garter snakes wouldn't take kindly to interference. But in late spring, Shine and his colleagues can do what they like with the reptiles – male, she-male or female. Pick them up, measure them, put them in a bag; the snakes don't have the energy to care, making them

almost ludicrously perfect for study. That's why Shine made
a pilgrimage from Australia to snake dens near Narcisse
seven years out of eight from 1997 to 2004. 'Having 10,000
amorous snakes in an area the size of a living room is a
snake biologist's idea of heaven,' he says.

To find out the she-males' secrets, Shine and his
colleagues simply sat in the grass alongside red-sided garter
snakes that were fresh from their winter quarters. Grabbing
individual she-males by the tail, the researchers presented
them to 'real' males to see how they'd react. The males
almost always found the she-male a turn-on, pressing their
chins on him/her and lining up their bodies. So males
definitely fall for the she-males' pheromone charms. But
what's in it for the she-males?

Time for a more cunning plan. Shine kept one group of
she-males at 10°C, the temperature of their bunker. He
warmed another batch of she-males to 28°C by putting
them in cloth bags and placing them on the electrically
heated front seats of the team's four-wheel-drive Yukon
hire car. Next the team brought the two groups to a
common temperature of 25°C, heating the cool snakes up
on the car seats, while letting the warm group chill off
naturally.

Holding each 25°C she-male by its tail, Shine presented
him/her to five different males. As expected, the males
flicked their tongues faster and tried to sidle up to the
she-male. But their interest didn't last forever. The guys
stopped sniffing around a snake from the 'warm' group
within about three hours. 'Cool' snakes won attention for
five hours. The males' loss of interest revealed that the
she-males had stopped gender-swapping and gone back to
being simple males, with the 'warm' she-males reverting to
type faster than the 'cool' ones. The conclusion was clear:
male red-sided garter snakes become she-males to warm
up as fast as they can. By pretending to be a female, a
she-male entices other males to press themselves against
what they see as a potential mate. Rubbing against his/her

warmer rivals, the cold she-male draws heat generated by their muscles into his/her own body. Heat, as we'll hear later, only ever flows from hot to cold.

Snakes and ladders

Stealing heat from a fellow animal is known as kleptothermy (not to be confused with a compulsion to race out of the supermarket with jars of coffee stuffed under your coat – that's kleptomania). As they're long and thin, snakes have a huge surface area for their volume, so they lose heat faster than if they were round and cuddly. Heat is a precious commodity in the chilly Manitoban spring; it's about 10°C if you're lucky at this time of year, the same as underground. By sliding against each other in a giant heap, snakes can cut how much heat they lose. It's like camping on a cold night: snuggling up to someone in your tent keeps you both warm.

Acting the she-male lets a recently emerged snake warm up pronto from his winter slow-down. It's vital to act quickly. Cold and sluggish after months underground, he's a target for crows, who'd love nothing better than eating a lethargic snake. Warming up will help him move fast enough to avoid the bird's clutches. Gender-swapping has more underhand benefits too; it distracts a she-male's rivals from the real females, making the males waste precious energy on unproductive relationships. Meanwhile, the she-male saves energy by not bothering to have sex. As the proverb goes, keep your friends close but your enemies closer.

Originally biologists thought only some male red-sided garter snakes have a she-male phase after brumation. It turns out they all do, though the gender-swapping doesn't last forever. After warming up for a day or two, most she-males revert to being male and head off on their summer travels. The bulk of the courting and mating between real males and females happens in small groups far away from the snakes' winter dens.

So 'absurdly easy to investigate' are the red-sided garter snakes in spring that Shine was able to knock out more than 40 scientific papers from those seven trips he made to the wilds of Canada. 'One can frame a novel idea one evening, test it the following day, and devise a follow-up experiment over dinner the next evening,' he says. With the snakes so obliging, perhaps Spielberg should have taken a leaf out of Shine's book and filmed that famous snake scene in *Raiders of the Lost Ark* up in Canada.

The heat is on

During this sizzling saga of snake skulduggery we blithely threw around terms like heat and temperature. With a bit of luck you hardly even blinked. Heat is a word we use every day. We talk about the heat from the Sun or being in the heat of an argument, and we all know what it's like to feel hot or cold. Yet even the biggest brains in physics once found it hard to understand what heat really is. Back in the eighteenth century, most scientists thought heat was an invisible, weightless fluid called caloric that slinked its way from a hot object to a colder one. Though we might laugh at the notion of caloric today, it took an experiment in 1798 involving animals – two horses, and a man studying cannon manufacture in Munich – to knock the idea on the head. American-born Brit Benjamin Thompson (1753–1814) made the horses walk round in circles, driving a metal drill bit so that it bored a hole in a 2.7kg (6lb) brass cylinder in a vat of water. After two and a half hours, by which time the horses must have been as bored as the cylinder, both the brass and the water were extremely hot. 'It would be difficult to describe the surprise and astonishment expressed in the countenances of the by-standers, on seeing so large a quantity of cold water heated, and actually made to boil, without any fire,' Thompson wrote.

Where did this heat come from? Rub your hands together and you'll get a clue. When two surfaces, such as two palms or a drill bit and a cannon-in-the-making, move against each other, they generate a force known as friction. This resists the movement and converts some of its kinetic energy – the energy of motion – into thermal energy, otherwise known as heat. Friction features again in the next chapter, along with some ancient Greeks and an ice-hockey puck. But the boring horses (anyone remember the old *Yellow Pages* category 'Boring, see Civil Engineers'?) didn't do all the work. By showing that the material properties of the boring rods, brass and water hadn't changed, and that the water warmed up for as long as the horses kept moving, Thompson proved that none of them had gained or lost caloric fluid. Yet heat had still transferred. Caloric fluid, although a handy explanation, didn't exist. Instead, Thompson thought heat was a form of motion, which, if you think of both concepts as types of energy, is true. But despite his and the horses' efforts, it took many other bright minds – including that of the Manchester-born James Prescott Joule (1818–89) – half a century to bury caloric theory once and for all.

Get a move on

Joule was manager of the family brewery but found himself sidetracked by science. He installed a lab and heated water with motion just as Thompson had done. No horsepower this time: he attached a weight to a string so that as it fell it swooshed a paddle wheel round in a tank of water. Joule was then able to calculate the mechanical work done by the falling weight. By measuring how much heat this work created in the water, he linked the mechanical energy needed to perform the work to the thermal energy it made. This experiment helped develop the principle of conservation of energy, which in a nutshell says that 'energy is neither created nor destroyed'. Instead it shifts from one form to another – in Joule's experiment, from mechanical to thermal. In a light bulb, the electrical energy of the flowing current turns into light and heat, while animals convert the chemical energy in their food into mechanical energy to move around.

The transfer of energy between heat and other forms is known as thermodynamics, with the first law of thermo-dynamics incorporating the principle of conservation of energy. There are four laws in total, and they are numbered – unconventionally – from zero to three. Physicists added the zeroth law of thermodynamics in the twentieth century after describing the first, second and third laws the century before. The numbering had to go backwards; the zeroth law is like a prequel. If the first, second and third laws of thermodynamics were *Star Wars*, *The Empire Strikes Back* and *The Return of the Jedi*, the zeroth law would be *The Phantom Menace* (or, if you're a *Star Wars* purist, *The Revenge of the Sith* – physicists, we need you to come up with a couple more laws of thermodynamics to make this analogy work properly).

As a result of his paddle wheel and other efforts, Joule gives his name to a unit of energy: one joule (J) is the energy transferred (or work done) to an object when a force of one newton (N) moves that object one metre. That's roughly the energy needed to lift a small apple a metre into the air (more on forces and newtons in the next chapter, including details of an excellent hairstyle and how to walk on the ceiling). After an absence of almost four decades, the Joule beer brand was relaunched in 2010 but its lack of physics-based ale names is disappointing (anyone for a pint of Dark Energy or Stellar Artois?). Joule's name is a familiar sight on food packaging too: one 25g (1oz) packet of salt-and-vinegar-flavour crisps that we ate for research purposes provided 540,000J (540kJ) of energy. That's the same as 130 food calories (where 1 food calorie, or strictly speaking, 1 kilocalorie, equals 4,184J). The calorie, named after caloric fluid, is out of date but calories still taste good even if they're not worth counting any more.

Today we define heat as a form of energy transfer. Going one better than caloric fluid, which doesn't exist at all, heat exists only when two objects are at different temperatures. Then it flows between them until they reach the same temperature or, in science speak, are in thermal equilibrium. At this point the energy transfer stops and the heat ceases to exist. The energy formerly known – and transferred – as heat is incorporated into the kinetic energy of the atoms or molecules jigging about inside the previously cooler objects, making them move faster. At all temperatures above absolute zero (–273.15°C or, in the geek's temperature scale of choice, 0 kelvin, or K), such particles jig about all the time. In a gas or liquid, they move about freely, while in a solid they vibrate around their 'fixed' positions. That's what temperature is: a measure of the average kinetic energy of the atoms or molecules in an object. Something at a higher temperature, with more energy in its molecules, can transfer heat to something that's cooler, like the ice cream you've just put on your hot apple pie. In other words, temperature is a measure of an object's ability to transfer heat.

But how does heat transfer work? Take those garter snakes. When a cold and lethargic she-male tricks, with the aid of his/her female-faking pheromones, 'real' males into huddling around him, how does their heat work its way out through their skin and into the she-male's body? It's not as if the snakes swap molecules, even if passing on their DNA is the duped males' number one aim. Instead, the snakes use conduction to steal body heat (in the physics sense; they're not waving a stick in front of an orchestra or checking train tickets). Where they're pressed up close against the cold she-male, faster-jiggling molecules in the warm 'real' males bash against the she-male's slower molecules. These collisions transfer some of the speedy molecules' energy to the neighbouring slowcoaches. For the she-male's molecules, it's like being jostled by a fast-moving crowd; they end up moving faster themselves. The resulting heat transfer lowers the average kinetic energy of the molecules in the real males, reducing their temperature, and boosts the average kinetic energy of the molecules in the she-male, raising his/her temperature. Conduction like this, with two objects at different temperatures in close contact, is just one of the ways of transferring heat; more on the other two soon. But first, let's go to the dogs …

Twist 'n' shake

… via one of humanity's greatest inventions: the hot bath. Perfect for idle contemplation and the occasional Archimedes-style 'Eureka!' moment. Showers might save water but they don't give you time to think. So there you are, lying in the bath with your favourite popular-science book, the scent of lavender wafting in the steam, Vivaldi in the background and a mug of peppermint tea on the corner of the tub. Bliss. You even manage to keep the pages dry when your mind wanders for a second and you wake with your mouth dipping below the surface. Hmm. Lavender soap doesn't taste as good as it smells.

Still, everything else is perfect. If you cool off, that's easily fixed. Just add more water from the hot tap and carry on doing nothing. The hot water plummets to the bottom because of gravity, then rises to the top as it's less dense – it has fewer molecules in a given volume than the rest of your now too-cool-for-comfort bathwater. As it gains height, the hot water pulls colder, denser water from the other end of the bath to the place it's just left. The result is a convection current that mixes everything up and distributes the heat without you having to lift a finger (though a quick swoosh with your hand works wonders even more quickly).

Convection – the second way of transferring heat – occurs in all liquids and gases because their atoms or molecules can move about freely. Conduction, in contrast, works best in solids, where the atoms or molecules are confined near set positions and tend to be closer together, although it can take place in liquids and gases too. So you can thank two types of physics for warming you up in the bath: convection moves the hot water to you and conduction gets the heat into your body, just as it warms a she-male garter snake in a mating ball. All's well again.

Eventually your fingertips become pale and wrinkled and you decide to get out. Disaster! As you heave yourself out of the tub, water dribbling down your body, you realise your towel is in the wicker laundry basket in your bedroom. The one your aunt gave you. Curses. It was toasty warm in the bath but now you're freezing as you scurry across the corridor, leaving sodden footprints on the carpet.

When you get out of a bath, as much as 0.5kg (1lb) of the liquid stays on your body – about 0.5 per cent of your total mass (please accept our apologies for commenting on your weight). In volume terms, that's roughly half a litre, or a small carton of milk. Most of the water runs off but what remains evaporates: the hottest, fastest-moving molecules escape through the surface of the liquid into the air, leaving cooler, more sluggish molecules behind. The average temperature of the remaining water falls, cooling you down.

Evaporation's handy in summer when you sweat, your body deliberately creating pools of water on your skin to cool you. It's also useful when a dog pants with his tongue out to let saliva evaporate from his mouth. But this physics phenomenon isn't ideal if you're towel-less and dripping dry in cold air after a bath. If there's a draught from a leaky window, you'll feel colder still as the moving air whisks the escapee water molecules away from the surface of the water droplet, giving others more chance of jumping out and hastening the evaporation. It's why climbing out of an open-air swimming pool on a windy day is so bracing. In the UK, at least.

Water on your skin also cools you by drawing thermal energy out of your body through conduction; water conducts heat about 25 times better than air as its molecules are nearer each other. Together, conduction and evaporation mean you'll feel much colder if there's a layer of water next to your skin than if there's only air. Grabbing a towel and drying yourself off as fast as you can is the only way forward.

Fur enough

We might have towels, even if we left them in the bedroom, but animals don't. A furry animal – be it dog, bear, panda or hamster – can trap a lot of water between the hairs of its coat. The bedraggled fur of a rat holds around 5 per cent of the animal's total mass as liquid. Scaled up to human terms, that would be like us having 4 or 5 litres (7–9 pints) of water on our skin when we step out of the bath, 10 times more than normal. But it could be worse. At least the rat's not an ant covered in tiny hairs that hang on to a whopping three times its body weight in water.

Having fur this wet could cause a serious drop in body temperature as evaporation does its work. Or if, like dogs, other mammals, birds and a few types of fish, the animal's an endotherm and so generates its own heat, then its sopping fur could cause its energy levels to plummet as it burns fuel to stay warm. Animals make big efforts to stay at the same temperature because their bodies work best only within certain limits. In endotherms, that range is generally just a couple of degrees Celsius (we're ignoring hibernation here). Reptiles, butterflies, moths and other ectotherms, which don't create much heat themselves, often cope with a wider span of body temperatures. Red-sided garter snakes, for example, are OK at the 10°C of a Canadian spring day but faster-moving and safer when their bodies are at 25°C. As a rule, ectotherms don't need to eat as much as endotherms, but there are downsides too. If ectotherms want to move far they must first lie around in the sunshine, they can't move fast for long, and they can't live anywhere too cold. Ectotherms also find it harder to be active at night (though geckos – see Chapter 2 – are nocturnal).

Enough of geckos, rats, pandas and ants – we're trying to talk about dogs. As mammals, they're endotherms and their bodies need to be about 38–39°C. Below 37°C or above 40°C and it's time to take your pooch to the vet (though don't rely on us for medical advice – only one of us has a first-aid certificate and it's out of date). Like us, if dogs are too hot, their metabolism (the rate their body burns food to release energy) speeds up and they use their resources too fast. Also, if the enzymes that enable those energy-releasing reactions get too warm, they will stop working. No energy means no life. Too far below their ideal temperature and those enzymes don't work well either. The dog's metabolism slows, along with its heart rate, breathing and brain activity. If the animal gets too cold, all these essential operations will stop.

The dog's body

Being wet is a double whammy: not only does it cool the dog through evaporation but it also prevents the animal's fur from keeping it warm. Normally, the hairs trap a layer of air, which conducts poorly and makes the coat more insulating so it loses less heat (the pockets of trapped air do set up convection currents but only over small distances). Humans do this too, even though we've lost most of our fur; when we're cold, we raise the hairs on our arms with goose bumps, trapping a thin layer of air next to our skin and cutting convection. Once a pooch's fur gets wet, however, it conducts body heat away faster because water has displaced the trapped air. More heat escapes into the air around the animal, forming larger convection currents, which take away yet more heat on top of that lost from evaporation. Brrrrr.

To summarise: dry dog fur minimises heat loss through conduction and convection. But if that fur is wet, the animal has to burn precious energy to stay warm enough for its body to work. No pooch is that daft, as you'll know to your soggy cost if you've stood next to a dog that's just bounded out of a river. Dripping wet, it shakes itself dry. By twisting its body from side to side, the dog spins the water off, sending droplets flying in all directions like a shower hose on the loose. It's not just dogs: all sorts of furry animals spin themselves dry.

Let's go for a spin

One day, David Hu of the Georgia Institute of Technology in the US was watching his dog, a toy poodle called Jerry, dry off when a couple of questions popped into his head. How much energy does an animal use to shake itself? And how much energy does this save by stopping evaporation from taking away body heat? Piqued by scientific curiosity, Hu, along with his students Andrew Dickerson and Zachary Mills, decided to find out why shaking is such an effective drying technique for dogs and other mammals.

Rather than setting up camp in the wilds to watch animals TV-documentary style, the three researchers videoed the shakes of 16 species sourced from Zoo Atlanta, Georgia Tech research labs and local city parks. The smallest was a baby mouse and the largest a brown bear, with everything from a rat to a squirrel, cat, kangaroo, lion and tiger in between. As well as Hu's poodle, the team looked at four other breeds of dog: a chihuahua, a chow, two Siberian huskies and four Labrador retrievers going by the names of Belle, Molly, 'I'm afraid we can't remember' and Chipper.

The tests were simple. The researchers sprinkled smaller animals, such as rats and mice, with water from a spray bottle and soaked larger ones with a hose. Then they filmed the animals spinning themselves dry, using cameras running at up to 1,000 frames per second, around 40 times faster than TV. The rats and mice looked the cutest; squinting their eyes, they lifted their pink front paws off the ground before twisting their bodies back and forth like the drum of a washing machine, going rapidly first one way then the other.

No matter what their size, all animals shake in roughly the same way, the team discovered. Not only does this spinning look sweet but it's also effective; it takes just seconds to get drier, even if not all the water comes off. The big difference between animals is in how fast each shakes its body. If you look at a spin-drying mouse head on and keep your eyes trained on a piece of fur at the top of its body, you'll see that spot move, say, clockwise, then come back to the middle before twisting anticlockwise by the same amount and returning to the start. The three researchers found that the tiniest animals generally make this oscillation back and forth the fastest, while the biggest go slowest. The mouse was the winning spinner, completing about 31 shake cycles a second. The brown bear – the largest animal in the study – was most lumbering, managing just four shakes a second. Domestic cats, at nine shakes a second, were solidly mid-table.

On a roll

What about man's best friend? Labrador retrievers famously appeared on the now-iconic 1972 TV advert for Andrex brand toilet paper (Cottonelle in the US) showing a sandy-coloured puppy gambolling round a house with the end of a loo roll in its mouth, trailing paper everywhere. Over 100 similar ads followed, the puppy's floppy ears and doe eyes proving a hit for sales. The dogs were also great for Hu, Dickerson and Mills, who examined their four retrievers in more detail than the other animals as they were so easy to work with – most of the time. 'You can always count on a dog to shake off, though not always facing the camera,' says Dickerson, who led the work.

After soaking the Labradors with water, Dickerson measured the frequency of the dogs' shake to be about 4.5 times a second, slower than a cat but in line with the size-versus-speed rule. Then he got more experimental. Dickerson taped a section of pink drinking straw onto the fur in the middle of a Labrador's back so he could measure how far – not just how fast – it rotated. The straw span round through an angle of about 90° to either side of the dog, he found. Water drops flew off throughout this cycle but did so most when the fur changed direction at either end of the spin. Like the other animals, the Labrador shook itself for several seconds before stopping.

Moving your fur round that far – 180° in total, or half a full rotation – is impressive. The dog's secret lies in the tissue between its outer skin and its muscles. This soft, spongy layer, made of collagen and elastic fibres, is particularly pronounced in dogs, protecting them from blows. By holding one of the Labradors still while moving its skin by hand, Dickerson was able to rotate this loose dermal tissue by up to 60° either side of the animal's backbone. The remaining 30° of motion to reach that 90° stretch comes from the twisting of the dog's backbone itself.

Shaken not stirred

The type of back-and-forth movement shown by all Dickerson's shaking animals, not just the Labradors, is common in nature. Known as 'simple harmonic motion', it also happens when a pendulum oscillates from side to side, or when a mass on a spring bobs up and down. The object that's moving – for a dog, its skin – goes fastest at the mid-point of the motion, before slowing to a halt at its furthest point from the centre, then speeding up as it moves back in.

We've known the mathematics describing simple harmonic motion for centuries; it's fundamentally the same whatever the situation. To estimate how much energy a Labrador retriever uses when shaking itself dry, all you have to do is tweak the standard formula for the maximum energy of an object moving like this. For one of Dickerson's

Labradors, that means multiplying the mass of all parts of the dog that are shaking (its muscles, skeleton, water, fur and organs) by the square of the radius of the dog's chest (about 12cm, 4.7in) and the square of the frequency of oscillation (4.5 times per second), then dividing the result by two. Then just multiply that figure by the total number of shakes the dog makes. The answer? Unfortunately, we can't calculate that precisely because we don't know how much of the dog is shaking. 'This equation gets you close, but isn't good enough for calculating an exact value, at least for a peer-reviewed [scientific] article,' Dickerson admits.

One big unknown is the fraction of its water that a Labrador flings off. To firm this up, you could try weighing a wet mutt before and after shaking, but it's not easy getting a soaking dog on a pair of scales. Dickerson's solution was to invent a 'robotic wet-dog-shake simulator'. This sounds fancy but was simply clumps of soggy dog hair clipped to a motor taken from a household electric drill. By whizzing the hair round, Dickerson discovered that roughly 70 per cent of the water eventually flies off, detaching as a series of drops. So if the dog has 500g of water on its body, it manages to fling off close to 350g, about the mass of liquid in a standard fizzy-drink can. No mean feat.

For a Labrador weighing 30kg (66lb), with 500g (1lb) of water in its fur, Dickerson reckons that, if the animal's skin didn't get in a flap, it would have to burn about 480kJ to replace the heat lost when 70 per cent of the water on its body evaporated, rather than being flung off. This 480kJ is equivalent to about 110 food calories. Given that a dog eats about 800 food calories a day, the wet Labrador would – if it didn't shake itself dry – have to burn roughly a seventh of its total daily intake to keep warm, assuming it converts all that food energy into heat energy. That's about a third of a can of dog food, which no mutt's keen to waste on staying at the same temperature. Dickerson is reluctant to put a figure on the energy the dog needs to shake itself dry, but he reckons it's about 100J, nearly 5,000 times smaller than that 480 kJ.

So loose skin and a twist will shake a dog dry. We humans don't have enough floppy tissue beneath our skin to do a good spin-dry – that's why we need towels. Our flap-free plight is shared by hairless guinea pigs. First bred in the 1980s, these 'skinny pigs' have a largely baby-smooth pink or brown surface, or sometimes a mix of both colours. Dickerson found their skin, like ours, is too tight to shake. Not having towels, they must simply shiver when they get wet.

Next time your dog leaps out of the sea and stands next to you poised to dry off, remember two things. First, it'll shake about 70 per cent of the water from its coat without moving away politely first. Second, the dog would need between 1,000 and 10,000 times more energy to evaporate the water than it needs to spin itself dry. It's one decision that doesn't need mulling over in the bath.

Radiate to accumulate

That's baths, wet dogs and two methods of heat transfer covered – conduction and convection. Before we look at how one insect avoids being killed by exactly the same hot beverage it needs to lay eggs, let's find out more about the third, and final, way of transmitting heat. We'll also discover why a wet black Lab chills off a fraction faster than a wet golden Lab.

Known as radiation, this method of heat transfer doesn't need atoms or molecules at all. The heat energy travels through air as electromagnetic waves made up of alternately pulsing electric and magnetic fields (which we'll come back to in Chapter 5). When the waves hit an object, they transfer energy to its atoms or molecules, making those particles move faster and warming the object up. That's why a garter snake basks in the Canadian sunshine – it absorbs electromagnetic waves from the Sun, which beam down to Earth through space, taking 8 minutes and 19 seconds to make the 150 million-kilometre (93 million-mile) trip.

You don't have to be a star like the Sun to radiate electromagnetic waves, however. You just have to be warmer than absolute zero (0 K or -273.15°C). And everything in the universe is; scientists have cooled small pieces of material to within a billionth of a degree of this temperature but not absolutely to zero (and they never could, no matter how hard they try; physics won't allow it). So all objects emit electromagnetic waves of one type or another as their molecules or atoms jiggle about. The exact wavelength depends on how hot the object is. Heat a poker in a log fire and you'll see it glow white-yellow as it warms and emits visible electromagnetic waves, or light. Take the poker out and, as it cools, it'll change colour from yellow to orange and then red as the wavelength of the light lengthens. Even when the poker reaches room temperature and looks like an ordinary bit of cold metal, it's still emitting waves, but they're infrared, which we can't see (handy, though, for your TV remote control).

Those changing electric and magnetic fields pulse at different frequencies, faster or slower, depending on the energy of the wave. To keep to the laws of physics, they must always travel at the same speed through a particular substance. This means that infrared waves, which contain less energy than visible red light and so have fields pulsing more slowly, must travel further each time the fields pulse; in other words, infrared wavelengths are longer than those of red light. Human eyes only see wavelengths from red to violet. More on this in Chapter 6, where we'll also find cuckoo chicks that use short-wavelength ultraviolet pigments to trick their foster-parents.

Like pokers, animals at room temperature give off infrared waves. We can't see these waves but physics tells us how they behave. Black objects look black because they absorb all the wavelengths of visible light and don't reflect any back to our eyes. That's why a beach with volcanic black sand is more likely to scorch the soles of your feet than one with golden sand – it's absorbed more light and so become hotter. Similarly, black objects emit electromagnetic radiation of all wavelengths – including infrared – most efficiently, which is why a wet black Labrador cools slightly quicker than a golden-haired dog.

Ultimately it's electromagnetic radiation – both visible and infrared – from the Sun that drives biology, providing heat at the right levels for the planet to sustain life, and light that helps plants to photosynthesise food for themselves and animals. While having enough heat is vital for life, in certain circumstances too much can be fatal. Let's turn to an animal that uses thermal physics to keep cool. It's an insect that can be deadly to humans. Even though, once you've heard his story, he's the last person you'd expect to love them, Manop Rattanarithikul is one of these creatures' biggest fans. It's time for a trip to Thailand to explore the life of the mosquito.

A nice hot drink

These days Rattanarithikul, together with his insect-specialist wife Rampa, is the proprietor of the Museum of World Insects and Natural Wonders in the city of Chiang Mai. Their museum boasts hundreds of species of mosquito displayed in wooden cases, along with numerous beetles, centipedes and leaf bugs. There are also garish paintings of giant mosquitoes that Manop has created himself, including one of a metre-long black specimen perched on a woman's lap in an emerald jungle. His liking for mosquitoes is surprising, because as a nine-year-old in Thailand during the Second World War he almost died of malaria, the killer disease spread by these insects. What's more, his treatment was the most painful you can imagine.

The young Rattanarithikul was living in Chiang Mai, where Japanese troops were stationed in a temple, when US pilots bombed the city. Keen to keep their son safe, his parents whisked him off to a village in the hills nearby. He may have been protected from the war, but it was here that Rattanarithikul contracted malaria. With no medical supplies available, he was taken to a local woman for a cure. According to a sign at the museum, she and her husband removed the boy's shorts 'and then got the thorn from the lemon tree and pierced the skin around my anus and pulled'. Rattanarithikul can't recall how often the couple did this barbaric act, but they continued until the blood poured from his bottom. 'The pain was excruciating,' Rattanarithikul says. 'The pain made me sweat all over my body.'

Crazy as it sounds, the idea was that the sweating would relieve the disease; the nine-year-old received this 'cure' a total of 10 times in two months. Somehow Rattanarithikul survived this agonising and pointless treatment. He made a full recovery, leading to a lifelong fascination with mosquitoes, and the museum. We're not telling this story to shock but because there's a curious parallel with mosquitoes themselves. After some mosquitoes have sucked blood from a human – or any other animal – they expel a bloody liquid from their own bottoms. Getting rid of the substance they've just drunk seems strange. There is a reason, however, and you won't be surprised to discover it involves physics.

Red red whine

Named after the Spanish for 'little fly', there are more than 3,500 species of mosquito; all of them make that familiar irritating whine by flapping their wings 500–600 times a second. In some, the body curves over near the 'shoulder' so that the head hangs down like a witch hunched over a cauldron. Mosquitoes generally feed on nectar and plant juices, but many species – though not all – suck blood too. It's only the females who want your blood, craving its proteins to help them lay eggs. Once deposited in water, the eggs hatch into larvae that later pupate, developing legs and wings. Eventually they emerge as fully fledged adults that rest on the water's surface until they're strong enough to fly. There'll be more about how some insects can walk on water in Chapter 3.

Adult mosquitoes live for no more than a couple of weeks and the female can't hang around. She gets the blood she needs by piercing her victim's skin with a long tube, or proboscis (from the Greek for 'forward-feeder') attached to her head. Videos shot in 2013 by Valerie Choumet and colleagues from the Pasteur Institute in Paris show that the proboscis isn't rigid like a drinking straw but buckles and bends to search for blood inside a wide area of flesh. When she tracks it down, the mozzie injects saliva that contains anticoagulants to stop the blood clotting and bunging up her proboscis, before sucking up the free-running liquid she desires. Sometimes she sucks so hard that the pierced blood vessels collapse, spilling blood into a pool that she can go back to for a refill. As Rudyard Kipling's 1911 poem put it: 'The female of the species is more deadly than the male.' Not, in the case of the mosquito, because of the small amount of blood she takes but, in many species, from the diseases she can transmit to you through her bite.

Female mosquitoes carrying malaria tend to bite between dusk and dawn; that's why you can wake up in the morning covered in itchy bites. The itching is due to your body's immune response to the saliva your female predator injected you with. More seriously, her saliva may contain parasites that cause malaria or the viruses responsible for yellow fever, Zika fever and dengue fever. According to data from the World Health Organization in 2013, almost 600,000 people die each year from malaria alone. Former Microsoft boss Bill Gates is one of those to have ploughed money into eradicating this disease. But if you think you can avoid bites by keeping your eyes peeled for female mosquitoes, you're wrong. It's impossible to tell the two sexes apart with the naked eye. With a microscope, though, you'd notice one big difference. A female's antennae are smooth, while a male's are covered in hairs he uses to pick up sound waves given off by the females' wings so he knows where to search for a mate.

Dangerous liaisons

Mosquitoes can be dangerous to humans, but these insects don't have it easy either. All it takes for a mosquito to be splatted into oblivion is a rolled-up newspaper or a deft flick of the hand. Even the Dalai Lama isn't averse to a spot of mosquito bashing. Interviewed by US journalist Bill Moyers, the spiritual leader seemed to suggest that if he can't blow or brush away a mosquito that's landed on him, he'll swat it instead. Should Buddhists kill mosquitoes whose bites may transmit deadly diseases? It's an interesting dilemma …

The biggest danger for a female mosquito isn't the Dalai Lama but the blood sucking itself. The blood is vital for her eggs yet the animal she's sucking it from is unlikely to be happy and will try to crush her as she dines. That's why mosquitoes prefer to bite at night – their victims are resting or asleep. Even then it's dangerous, as the blood 'donor' could wake at any time. Keen to avoid trouble, the female drinks as much blood as she can as fast as possible. It's like breaking into an all-you-can-eat restaurant where the owner will kill you if they catch you guzzling the hot soup. No point hanging around. The female mosquito gorges herself silly on her liquid lunch, typically trebling her body weight by taking several milligrams of blood in one feed. For us to triple in mass, we'd need to wolf down well over 100 litres of soup in a single sitting.

This bingeing makes sense if it keeps you alive. Trouble is, humans and other mammals have warm bodies, usually around the 34–40°C mark. So when a mosquito sucks up our blood, heat rushes into her too. That's a problem, as she may become too warm to reproduce and her life is too short to waste time. Bloated and hot, the female may even become a victim of blood-sucking herself, with another female plunging her proboscis into her rival's freshly acquired stock of blood. This isn't that unlikely; mosquitoes are good at hunting, either by eye or with sensors on their antennae that detect carbon dioxide from their quarry's breath. When a female detects another mozzie chock-full of hot blood, it's hugely tempting to take a bite.

The mosquito is stuck between a rock and a hard – or rather, a hot – place. So what's a girl to do? If she drinks slowly to avoid becoming too hot too fast, her victim may turn killer and squash her. But if she drinks quickly to avoid flailing limbs, tails or newspapers, she'll overheat, her body will malfunction and she might be sucked dry by one of her 'sisters'.

For the answer to how the female solves her problem, we can thank Claudio Lazzari and Chloé Lahondère from the Université François-Rabelais in Tours, France. After exploring how another insect – a flat-backed blood-sucker called a kissing bug that transmits Chagas disease – finds its prey by detecting body heat, in 2012 the pair installed a toolbox-sized cage full of *Anopheles stephensi* mosquitoes in their lab. Found in many parts of India, South-east Asia and the Middle East, these insects are one of the key culprits for transmitting malaria, so it's important to know more about their habits. Lazzari and Lahondère let the mosquitoes gorge on human hands or hairless mice, which both have smooth skin that makes it easier to see what's going on. Selflessly, the two researchers used their own hands. 'Both Chloé and I got bitten about 30 times,' says Lazzari. Just as well their mosquitoes were disease-free.

Biologists have known since the 1930s that female mosquitoes push out droplets containing blood from their bottoms. But Lazzari and Lahondère were the first to use a thermal camera to measure the infrared radiation from the insects as the blood comes out. The result was vibrantly coloured 'heat maps', showing red for hot, yellow for warm, and green and blue for cooler. The images resemble Andy Warhol's paintings of Marilyn Monroe, with Marilyn's face replaced by a side-on profile of a mosquito piercing its victim's skin. Taken 5 seconds apart, they chart the precise temperatures across a female *Anopheles stephensi* before, during and after sucking blood from hands held at temperatures from 28°C to 37°C.

In the first heat map of a 36°C hand, you can see the mosquito's head is as hot as the hand: it shows up bright red. Her body's a little cooler at 34°C (plotted in orange) although still way above her comfort zone of 25°C, which would show up blue. About 5 seconds after the female starts sucking blood, a sphere of liquid – a 50/50 mix of blood and urine – emerges from her bottom. Coloured green on the heat map, the droplet is initially at 31°C. Over the next 25 seconds it grows larger and cools to 26°C, turning blue in the thermal image. Then the sphere falls off. Nothing daunted, the mosquito squeezes out another droplet straight away.

For the 25 seconds that the sphere hangs on, the heat maps revealed that the mosquito's body temperature (rather than that of the droplet) falls by about three degrees, to roughly 31°C. Heat transfer begins within seconds of the drop forming. When she finally finishes drinking and pulls her proboscis out of her victim, the insect's body cools to the temperature of the air in the cage, 23°C. To confirm that the females heat up because of the blood they ingest, rather than through conduction from the hand they're sitting on, Lazzari and Lahondère turned to the opposite sex. When male mosquitoes land on a hand, they stay at room temperature, the researchers discovered, proving that it's blood-sucking that warms the females.

By mapping a female mozzie's temperature as she drinks, Lazzari and Lahondère proved that she expels that droplet of wee and blood to cool herself down and avoid heat stress. Just as we feel cold when we're fresh from the bath and have water drops sitting on our skin, water evaporating from the droplet on the mosquito's bum makes her cooler too. She takes advantage of evaporative cooling, just as we do when we sweat. What's more, as some parts of her are warmer than others, she's set up a temperature gradient that encourages more heat to flow from her hot regions to her cold ones.

Serve chilled

Mosquitoes aren't alone in cooling off by expelling drops of liquid. While buzzing through the air on hot days, a honeybee will spit out nectar from its mouth to stop its brain from overheating, allowing it to fly at up to 46°C. Some moths dribble fluid onto their proboscis to cool their heads, while greenfly (the enemy of any tomato-grower) excrete honeydew − a sugar-rich sticky fluid − through their bottoms to chill their bodies.

Before we leave mosquitoes, there's just one more thing, as detective Columbo used to say in the 1970s TV series. Why does the female mosquito include blood as well as urine in her cooling droplets? Surely this wastes food and makes her spend longer sucking on her prey, risking her life? Lazzari thinks that adding blood lets the mosquito expel a larger droplet. Because bigger drops have a greater surface area, they speed up the rate of evaporation and cooling. Squirting out blood also means the mosquito weighs less and can fly off faster if things turn dangerous. Mystery solved. Not every species of mosquito squeezes out droplets of blood, though. Lazzari and Lahondère discovered that *Aedes aegypti*, which spreads yellow fever, dengue fever and Zika fever, shows no signs of blood-excretion. Rather than cooling themselves down as they feed, these mosquitoes create molecules called 'heat-shock proteins' that repair the damage to any cells that get too hot.

But for *Anopheles stephensi*, the physics of heat flow makes females from one of the most deadly species of mosquito safer while they suck our blood. Although that knowledge won't lead to a cure for malaria, it's another step in the quest to understand how these insects spread the disease. It's valuable work but – in contrast to Manop Rattanarithikul – Lazzari doesn't even like mosquitoes. When he's not in his lab, he admits to hating these insects. 'But as a scientist,' he says, 'I find mosquitoes fascinating.'

Sting in the tail

Heat isn't only about being at the right temperature. Some animals, from Japanese bees to Californian squirrels, use this form of energy transfer more ingeniously, to defend themselves, or their young, against attack. Let's start with the bee. It's near the beginning of the alphabet and there's nothing physicists like more than being systematic. Apart from equations. And lab coats. Let's move on.

Life as a Japanese giant hornet (*Vespa mandarinia japonica*) could feel invincible. The insect is 5cm (2in) long, about the size of a little finger, and has a sting that kills 30 to 50 people in Japan each year. Its body, horizontally banded with shiny yellow-orange and black stripes, has an outer coating, or cuticle, that's too hard for the stings of the bees to penetrate. But the hornet is not all brawn. It also has a nifty system for attacking the European honeybees (*Apis mellifera*) imported into Japan for their high honey yield.

First the Japanese giant hornet sends out a scout hornet to track down a promising-looking beehive or bees' nest and nab a few bees. This envoy rips off the bees' heads and legs with her powerful mandibles (jaws) and brings their juicy, nutrient-rich torsos back to the hornets' nest to feed the larvae. After a couple of return trips, the scout heads back to spray pheromones (like the signals used by garter snakes to reveal – or fake – their gender) onto the bees' home. Other hornets pick up the scent and assemble at the beehive. They hunt as individuals until at least three are in position. Then they attack together, in what biologists have bluntly dubbed the slaughter phase.

The name is appropriate – each hornet kills around 40 of the European honeymakers per minute. Once teamed up in a group of 20 or 30, the hornets can dispose of an entire hive of 30,000 European honeybees in three hours; they're unique in the hornet world for working together on a kill like this. Next comes the occupation phase, as the group takes over the hive for 10 days, plundering the 'food larder' full of the European bees' orphaned and defenceless larvae and pupae (unhatched adults), and taking them back to their own nest to feed their young. Job done.

But there's one bee that has evolved a cunning way of fighting back. Meet the Japanese honeybee (*Apis cerana japonica*), which has had to deal with the Japanese giant hornet for longer than its European cousin. Its sting can't get under the hornet's exoskeleton, but the bee has strength in numbers, particularly against that first lone hornet scout. As Masato Ono, Masami Sasaki and colleagues at Tamagawa University in Tokyo discovered, the Japanese honeybee uses those numbers, along with physics and a spot of physiology, to team up against its enemy. Crucially, there's a small difference between one aspect of their bodies, something that would normally be irrelevant. By behaving in a certain way, the Japanese honeybees create the right conditions for their one superior power to defeat the hornet. You could say they make their own luck.

For the scout, it's like David versus Goliath only with 500 Davids. In the photos from Ono's paper in the journal *Nature*, the honeybees are barely as long as the hornet's orange face. The bees are small but they have a secret weapon as powerful as David's catapult (minus the stone-lobbing). Annoyingly, and this isn't even the cleverest bit, the bees pick up on the scout's 'here's dinner' pheromone marker and use it as an early warning signal, making the chemical a kairomone (a message that benefits the animal receiving it but harms the sender). When they smell and see the scout at their nest one autumn, around 100 worker Japanese honeybees crawl around the nest entrance. As the visitor nears, these guard bees lift and shake their abdomens, as if brandishing a fist to make a bully back off. Then the workers return meekly inside. The scout is set on attack and doesn't realise this is a trap.

As soon as she crawls into the nest, the scout meets an ambush. A thousand worker bees have abandoned their honeycomb duties to defend their home. Up to 500 of them cluster around the hapless hornet in a tightly packed bee ball just a hand's span across. After 20 minutes of this sinister group hug, the bees disperse and the scout lies dead, alongside two or three of the honeybees. Guard bees drag the corpses away and all evidence of the hornet's invasion attempt is gone.

Biologists used to think the Japanese honeybees sting the hornet to death. Turns out, though, their stings aren't up to the job. Fortunately, this cuts both ways: the bees cluster round the hornet so quickly and tightly that she can't sting either and her only defence is to bite. So how does the hornet at the heart of this great ball of bees perish, if not through death by 500 stings? And, if conditions in the middle of the ball are that bad, how come the bees right next to the hornet, apart from the few she manages to bite, survive?

To find out more, in 1995 Ono and colleagues recorded the temperatures inside the bee ball with infrared cameras. The centre reached a toasty 47°C, their heat-map images showed. The pictures render the ball of bees as a psychedelic cauliflower head, with fluorescent pink and red patches highlighting the warmest spots in a sea of yellow on a purple background. Even the 100-odd bees that grouped themselves outside the nest entrance in response to the hornet scout's chemical signal heat themselves up. Their bodies show as lone yellow splodges, with pink centres bordered with red; the temperature of each bee's thorax, or chest, is much higher than in unagitated bees outside the nest.

You shall go to the ball

So how do the bees, which are ectotherms and rely on external sources of heat, warm things up? Their technique is simple: they vibrate the muscles in their thorax, which, as well as moving, generate heat. You'll have noticed this if you've been for a run on a cold day. Within minutes of setting off, you were no doubt much warmer and taking off your woolly hat. Together the hundreds of bees in the defensive bee ball generate a lot of heat; and since their bodies – like those of the red-sided garter snakes in their winter pit – are so tightly packed, this heat is likely to pass from bee body to bee body, and on to the hornet, via conduction, perhaps with a spot of convection and radiation in the few small air gaps. At 47°C, the ball's peak temperature is about the summertime average in Death Valley. For a human that's an unpleasant 10°C hotter than our body temperature. If our heat-regulation systems break down and our temperature rises to 40°C or more, we'll suffer heatstroke and die. It can't be much fun for the bees either. Biologists know they can cope with temperatures up to 48–50°C, so 47°C is near the top of their range. Even the Sahara Desert ant (*Cataglyphis bicolor*) – one of the most heat-tolerant animals there is – can't deal with temperatures much more than 50°C (see Chapter 6 for more on this insect's amazing way-finding). For the giant Japanese hornet, it's game over. It only survives up to 44–46°C, a bodily limit that gives the bee the edge and means the bee ball fells the hornet with both physics and physiology.

This difference in heat tolerance may have arisen because the bees survive the winter together as a colony, producing and sharing heat to keep warm. Over time this may have upped their heat tolerance. In contrast, only young hornet queens over-winter; the other hornets live for just one season, dying off in the autumn and leaving the lone queen to start a colony afresh in the spring by laying her fertilised eggs.

Not only is a hornet at the centre of a bee ball too hot, but she's also surrounded by the carbon dioxide breathed out by all the bees. She definitely overheats and probably suffocates too, although the extra carbon dioxide may simply make her less able to cope with heat. Either way, after about 20 minutes the bees break away from the ball and disperse, their victory complete.

Social whirl

For the giant Japanese hornet, it's a case of 'if you can't stand the heat, stay out of the bees' nest'. But why does this insect raid Japanese honeybee homes if she's going to end up baked alive? And why does she dice with this unpleasant death only in the autumn? The answer, as so often in biology, lies in reproduction. At this time of year, the hornets are raising larvae destined to be the next generation of reproducers – the new queens and fertile males. These larvae need more protein than their sister larvae that will turn into bog-standard worker hornets. So the current workers are under pressure to provide lots of top-quality nosh. It doesn't help that the large caterpillars and beetles they caught earlier in the season are now in short supply. Needing nutrients, the hornets risk more dangerous food-gathering, attacking the colonies of Japanese honeybees and other smaller hornets. For failure isn't guaranteed – if a few hornets manage to reach a small colony before the Japanese honeybees kill the first scouts, the bees may flee, saving themselves but leaving their larvae behind for the hornets to take home and feed to their young.

So the defensive ball formed by the Japanese honeybee doesn't always work, even though it probably evolved in response to the Japanese giant hornet's bully-boy tactic of attacking in a group (it's the only hornet known to hunt in a pack against other social bees and wasps). The Japanese giant hornet, which as a lone scout was fighting 500 small Davids, ups its own odds by teaming up with its nestmates to create 30 Goliaths. European honeybees, which didn't have to deal with this team attack until we transplanted them to Japan, aren't nearly as good at forming defensive balls. They do their best to sting the scout hornet, and, while their stings aren't strong enough to pierce her cuticle, they may get lucky and penetrate the membrane between the segments of her thorax or around her wing joints. But trying to sting means they can't form as tight a ball around their attacker, limiting how much heat they generate and their ability to fry her. This generally means the hornets are able to gang up and destroy the whole European honeybee colony. Ono and colleagues also found that European honeybees don't react to the hornet scout's marking pheromone; they're way behind the Japanese honeybee in its evolutionary arms race with the Japanese giant hornet. The Japanese honeybee is often too hot for the hornet to handle.

A tail tale

As we've seen, garter snakes, dogs and mosquitoes use heat-related physics tricks to defend themselves against temperature change; they don't want to get too cold or too hot. As for Japanese honeybees, they use heat to protect themselves from the fearsome bite of the Japanese giant hornet. But what if your enemy's using thermal physics to hunt you down? Read on to find out how a humble rodent uses a classic military tactic, deceiving and confusing its attacker to defend itself. We're talking squirrels. Not grey squirrels (*Sciurus carolinensis*), the kind Ronald Reagan fed acorns to on the White House lawns in the 1980s. Nor the red squirrel (*Sciurus vulgaris*) native to Britain and immortalised by children's writer Beatrix Potter as Squirrel Nutkin but squeezed out by grey squirrels shipped across the Atlantic by rich Victorian landowners keen to see 'exotic' creatures on their estates.

Unlike tree-dwelling red and grey squirrels, our squirrel keeps its feet on the ground. In California, to be precise, so no prizes for guessing its common name. Found throughout the Golden State and up into Oregon and Washington, the California ground squirrel (*Otospermophilus beecheyi*) digs itself burrows, where it hangs out with its family. The squirrel – one of more than 300 species – has a mottled, grey-brown body and a sandy-grey underside. White fur surrounds its black eyes, while the hairs edging its ears are black. At 15cm (6in), the squirrel's bushy tail is about half the length of its body and key to our story. For this mammal has a nifty heat-related trick up its sleeve or, rather, up its tail.

The squirrel's a homebody at heart. When it does venture outside its burrow, seeking seeds and small insects to eat, the animal rarely strays beyond 25m (80ft) from home. This sounds dull, but our squirrel's not dumb. It plays safe as it's a target for eagles, raccoons, foxes and badgers. Enemy number one, though, is the rattlesnake, which loves wolfing down baby squirrels. Yet the battle between snake and squirrel isn't as one-sided as you'd think.

Snake, rattle and roll

A staple of cowboy movies with plots that need a sudden injection of drama, the rattlesnake is not a reptile to mess with. When it spots potential prey, be it a mouse, bird or California ground squirrel, the snake slithers along until it's close enough to lunge at the animal with its fangs. As it bites, the reptile squeezes muscles on poison glands near the back of its head to send lethal venom through its choppers, and on into its victim. The venom stuns the prey, attacks its tissues, and kills it in minutes. After swallowing its dead victim head-first, the rattlesnake slides to a quiet spot to digest the whole body with its strong gastric juices.

Living throughout North and South America, the 36 species of rattlesnake deserve our respect too. Get bitten and you could die, which is bad news for cowboys or anyone who thinks it's a good idea to pick up a rattlesnake and take a selfie. We're not kidding: a Californian man tried just that in 2015 and landed a $150,000 hospital bill after the unhappy snake attacked him. In most cases, if you take anti-venom within two hours of a rattlesnake bite, you'll probably pull through. If anything, humans are more of a threat to the snakes than they are to us. One danger comes from rattlesnake rodeos – supposedly 'fun' events held in the Midwest and southern US where people round up, display, eat and sell these reptiles. Other animals are a menace as well, from ravens and raccoons to skunks and weasels, which all like to prey on rattlesnakes, especially the tasty young ones. Then there's the kingsnake (*Lampropeltis getula*), a constrictor that's immune to rattlesnake venom and will swallow its rival head-first.

To protect itself from these enemies, a rattlesnake's tail is made of 10–20 hollow rings of hard keratin protein, the kind that's also found in human fingernails. By contracting muscles at the end of its tail, which is usually paler than the rest of its body, the snake knocks the rings against each other, creating a characteristic rattling noise that warns predators to stay away or risk a poisonous bite. So important is this rattle for self-defence that the snake protects its tail by raising it off the ground while moving.

Slide to it

The species we're interested in is the Pacific rattlesnake (*Crotalus oreganus*), one of the seven species native to California. About a metre (3ft) long, it's brown or grey with a line of white-ringed darker blotches, like a diamond-patterned pair of tights. With its venom primed for action, surely the snake will have the upper hand over any squirrel? Fortunately for the California ground squirrel, this rodent is tougher than it looks. The two animals have been at loggerheads for so long – living alongside each other for millions of years, according to the fossil record – that some California ground squirrels have become immune to rattlesnake venom. A female squirrel will even chew on the skins shed by these snakes before licking herself and her pups to disguise their scent.

If a California ground squirrel does encounter a rattlesnake, it draws on an arsenal of defences to deter its enemy. It may kick sand in the air to provoke the snake to rattle its tail. Acting the (beach) bully might seem foolhardy, but by annoying the snake, the squirrel susses out how big and active its opponent is. The rodent will also hoist its bushy, grey-brown tail to a vertical position, sweeping it from side to side a couple of times a second, like someone waving farewell with a hanky. Then the squirrel will pause, examine the predator, move closer and 'flag' its tail again.

Look online and you'll find some amazing videos of squirrels and rattlesnakes. In one encounter, a squirrel darts towards a snake and sinks its teeth into the reptile, which fights back, forcing the squirrel to retreat. The rodent jumps in for another bite, finally leaving the snake for dead. One-nil to the squirrel. In another, it's a win for the snake, which eats the entire rodent. In a third fight, the two animals square up and there's a long stand-off before the snake slinks away. It's a draw, but we'll mark it down as a victory for the squirrel as it's seen off its predator.

Researchers used to think the California ground squirrel shakes its tail to warn other squirrels and the snake. We're anthropomorphising again, but if squirrels could talk, the California ground squirrel would be saying 'I'm big and I've seen you. I'm gonna give you grief unless you clear off and look somewhere else for a meal.' Thanks to Aaron Rundus of West Chester University in Pennsylvania, US, however, we now know there's more to this tail-swishing than meets the eye. The squirrel has a secret. And guess what? Tools from physics, with the help of a set-up fight and a biorobotic squirrel (what would Beatrix Potter have made of that?), have revealed what's going on.

Showdown time

Back in 2007 when he was at the University of California, Davis, Rundus set up what has to be the ultimate squirrel/snake showdown. If this were a wrestling match, we'd put on a comedy announcer voice: 'I-i-i-i-in the sandy-grey corner, all the way from Winters, 100km north-east of San Francisco, a California ground squirrel. And i-i-i-i-in the brown corner, make some noise for a Pacific rattlesnake from California's Central Valley. Ladies and gentlemen, let's get re-e-eady to rumble.' The crowd goes wild.

Rundus didn't just let the two scrap it out in the middle of his lab. That wouldn't have been scientific. Instead, he built a cage as big as a two-drawer filing cabinet, with two of the vertical sides made from wire mesh, the other two from wood. That was the squirrel house. Then he made a larger 'testing chamber', the size of a toddler's play-pen. The rattlesnake didn't get its own house; it had to make do with a pyramid-shaped mesh cage in one corner of the testing chamber. Game on.

Once the squirrel had settled into its new home, Rundus put the squirrel house next to the testing chamber and opened a small door between the two. He wanted to see what would happen when the squirrel entered the chamber. Like some wrestling matches, the fight wasn't for real. The snake was safely enclosed the whole time; it was cage-fighting, but not as we know it. Still, the squirrel could see, smell and hear the snake. In a typical encounter, it approached the reptile cautiously, flagging its fluffy tail, sometimes trying to throw sand despite there not being any on the floor. Rundus let each experiment run for about 10 minutes before retiring the squirrel.

Keen to analyse the bouts at his leisure, Rundus captured them on film. Rather than recording with an ordinary video camera, he used one that measures infrared radiation. That might seem odd, but rattlesnakes are pit vipers. Members of this family don't just have eyes that see visible light; they also have a pair of heat-sensitive pits that pick up the infrared waves that most animals, including humans, can't see (there's more coming up on an insect that 'hears' infrared). Lying on either side of the snake's head, halfway between and slightly below its eye and nostril, these pits are millimetre-sized holes in the skin that lead into a hollow chamber.

The pits aren't entirely hollow, however. Stretched vertically across the middle of each chamber is a thin membrane. Any infrared radiation striking the membrane triggers nerves that send signals to the snake's brain. The size of the signal depends on the temperature of the object, such as a squirrel tail, emitting the infrared – it's big if the object's warm and small if it's cool. Thanks to its pits, the snake builds up a thermal map of its surroundings. Combining the data with visual information from its eyes gives the snake an extra-powerful sense of what's going on nearby. It's like having an in-built infrared camera as well as a 'normal' one. We don't know how the world appears to the reptile, but this ability to visualise warm objects is a handy skill, particularly in the dark of night, when animals with ordinary eyes can't see well. Armed with its heat data, the rattlesnake can creep up on its prey unannounced. A remarkable, but not a friendly, face.

Flagging the danger

Mind you, the California ground squirrel has been in this 'wrestling ring' before and is well aware that the snake has heat-seeking super-powers. The rodent has developed a cunning counter-measure. When a squirrel confronts a caged rattlesnake and flags its tail, Rundus discovered, the bushy appendage changes colour on the camera's heat maps, going from purple (23°C) to red (25°C). The higher temperature shortens the wavelength of the infrared radiation that the squirrel's tail emits, and boosts how much of it there is too. That's what the camera picks up. The rodent, it seems, is warming its tail by 2–3°C above the background temperature by pumping in blood.

The squirrel knows that the snake, with its fancy pits, will sense something odd going on, so the rodent deliberately sets out to confuse the enemy with a strong infrared signal from its tail. It's a perfect counter-attack. Rattlesnakes know that a tail-flagging squirrel is a nuisance – no-one wants to be bitten to death – so a hot tail puts the snake on the defensive.

But how do we know the squirrel performs its warm-up act because of the rattlesnake? Perhaps it's just heating its tail on a whim. To double-check, Rundus set up another cage fight, this time between a California ground squirrel and a Pacific gopher snake (*Pituophis catenifer catenifer*). Found in the south-east US, this non-venomous reptile looks like a rattlesnake – it's sandy with brown-black blotches – but its head is more rounded and overall it's longer. Like the rattlesnake, it also has a penchant for squirrel pups. The difference is that a Pacific gopher snake doesn't have pits on its face. It can't detect infrared and so has no heat-seeking powers. When a squirrel sees a Pacific gopher snake, Rundus realised, the animal raises its tail to make itself look bigger but doesn't heat it up. In a nutshell, the squirrel warms its tail only when infrared-sensing rattlesnakes are around.

Squirrelling away

So the squirrel changes its behaviour depending on the kind of snake it confronts: hot tail for a rattlesnake, cool for a Pacific gopher snake. But let's look at things from the rattlesnake's point of view. How do we know it's deterred by 'seeing' extra infrared radiation beaming from a squirrel tail? Perhaps the reptile would slither off even if the squirrel waved a tail the same temperature as the rest of its body.

Rundus couldn't ask a squirrel if it would mind keeping its tail cool. Instead he did the next best thing and built a biorobotic squirrel – a stuffed squirrel with a set of cylindrical cartridge heaters up its tail. Usually these devices are employed in industry for warming things like metal pipes, not squirrels, which raised eyebrows when Rundus went shopping. 'I got a lot of interest – and laughs – when the companies found out what I was using their heaters for,' he recalls.

With his biorobotic squirrel positioned next to some snake food so it looked like a real squirrel defending her pups, Rundus set the rattlesnake free from its cage. When the snake spotted a biorobot with the heaters turned off, it went on the attack, raising and drawing back its head ready to strike. But when the heaters were turned on, mimicking a live squirrel that's warmed up its tail with extra blood, the rattlesnake was much more cautious. The reptile spent less time in the test chamber and more inside its cage. It rattled furiously as it coiled itself, ending up with its head and neck cocked back, poised to strike, and its rattle held vertically, sticking out from beneath the coils of its body. Classic rattlesnake defensive manoeuvres.

What goes through the snake's mind when it sees a strongly infrared-emitting squirrel tail waving about? 'That's the million-dollar question,' admits Rundus, who reckons a hot squirrel tail could mess with the snake's head in one of two ways. Either it confuses the snake by creating an unusual amount of infrared activity in its heat-sensing pits. Or, as we said earlier, and this might be more likely, a hot tail that's waving makes the squirrel extra conspicuous, especially at dusk or night-time when light levels are low. It's warning the snake it might be in for a nasty time.

With the aid of his biorobot and some cage-fighting, Rundus proved that the California ground squirrel knows what to do when it sees a rattlesnake on the prowl: get its blood pumping. A hot tail fazes the reptile with radiation physics, forcing it to back off from its feisty opponent. The snake transforms from an effective predator into a wimp. But is the rattlesnake fighting back and developing its own counter-counter-strategy? Rundus thinks that's possible. 'There is some anecdotal evidence that snakes gauge how defensively a squirrel is acting when deciding whether to search burrows for pups,' he says. 'Moms with pups will act more defensive than females without.' While rattlesnakes are wary of tail-flagging squirrels, some may be overcoming their fear to help them find baby squirrels. The game of snakes and flaggers goes on.

Beetle drive

Let's stay in California. There's more to the state than deserts filled with ground squirrel/rattlesnake stand-offs. We're heading to California Memorial Stadium, Berkeley, near San Francisco. It's 1943, a normal day at the football game; the California Golden Bears playing at home against the USC Trojans, who are on a winning streak. The crowd are out to have fun, laughing and smoking in the California sunshine as they wait for the game to begin, the smell of hot dogs searing through the air, and the sound system crackling as the stadium announcer shouts everybody to fever pitch, his voice competing with the cheerleaders' chants. Nobody's bothered about the Hayward Fault, which runs directly underneath the stadium and could start an earthquake at any moment. The Bears need a win; their side of the stadium is a sea of blue and gold.

'Hey, cut that out,' comes a voice from the crowd.

'What?' yells the man behind.

'Flicking ash at my neck. What's wrong with you?' The first guy cranes his neck behind him as he speaks, his cigarette wobbling in the corner of his mouth.

'Excuse me?' The guy on the row above has shoulders so wide that the seams of his shirt are straining. The men next to him edge away into the crush.

'Ow!' Another man that time. 'Whoa, look at that.' He holds up something small and black between his thumb and forefinger. 'Little ★★★★★★ bit me on the hand.'

The language at those twentieth-century ball games was disgraceful. But in his defence, the guy was under stress. All around him the match-goers were swatting away beetles, their 20,000 cigarettes flying dangerously close to their neighbours' faces.

What was going on in this only slightly dramatised scene? The aggressors were two types of fire beetle, *Melanophila consputa* and *Melanophila acuminata*, which swarmed to the stadium from the pine-covered hills nearby. Beetles don't usually attack people, especially *en masse*. Were they unhappy with the score?

The answer is almost as strange as what happened: trees are to blame. Trees and, of course, physics. That stuff gets everywhere. In a way, the beetles were acting in self-defence. Not against the football fans, who they probably bit for food or in response to being squashed, but against trees. The beetles have developed an unusual lifestyle to protect their larvae from this unlikely source of attack. Trees, it turns out, though not famous for their aggressive qualities, pick on young beetles-to-be. They'd probably claim it was self-defence too, as the larvae try to munch through their trunks. The trees have a number of tactics; they spew out sap and resin, or make their cells replicate at top speed. 'These tiny larvae that you can hardly see with your bare eyes are squeezed by the dividing cambium cells and finally killed,' says Helmut Schmitz, a fire-beetle expert at the University of Bonn in Germany. The trees' bark is literally worse than the larvae's bite.

For a fire beetle, the only good tree is a dead tree. Trees that are still alive defend themselves; trees that are dead can't fight back. But how can an insect that's just 1cm (less than half an inch) long kill a tree towering metres above? To find out, we should examine the insect for super-powers. Fire beetles from the genus *Melanophila* live on every continent except Antarctica and Australia, where the 'Australian fire beetle' *Merimna atrata* has taken up the same survival strategy. Since the *Melanophila* genus and *Merimna atrata* are part of the jewel beetle family, Buprestidae, you'd expect fire beetles to have bright iridescent colours that change depending on the angle you look at them from, like a soap bubble or a security hologram. Somewhere along the way, however, the 11 species of *Melanophila* fire beetle lost their iridescence. Now they're charcoal black. They may not be pretty but at least they're safe from beetle collectors. This wasn't why they changed colour; collectors haven't been around long enough to alter evolution much. The fire beetles' dull appearance, as well as making them the plain Janes – and Shanes – of the jewel beetle family, camouflages them, keeping them safe from predators like birds and lizards.

Light my fire

So it's not the beetles' looks that kill. They use their colour to hide, not attack. Just how are these black insects bugging their tree enemies? By bringing in reinforcements, that's how. The beetles turn to nature and physics for help, exploiting the power of lightning. Disappointingly, the beetles don't crawl up each tree, tie a metal point to the top and stand back waiting for lightning to strike, cackling like a James Bond villain. Instead, they let nature take its course. Sooner or later lightning strikes without the beetles' help, starting a forest fire that kills trees across a wide area. Fires begin in other ways too; the beetles aren't fussy, they take advantage of those as well. Whatever the cause, once their enemy's bark is scorched, the fire beetles move in, laying eggs in the damaged, defenceless wood beneath. When the larvae hatch, they don't mind that their dinner is burnt; at least it's not trying to drown or squash them.

But the fire beetle's problems aren't over yet. If you don't start the burning yourself, finding a fire can be hard. For one thing, in the northern forests where many of the beetles live, there's a fire only every 50 to 200 years or so. That means the insects may have to travel a long way to find a safe home for their young. And if there is a blaze, the beetles are active during the day so they won't see its glow on the night horizon. They also have to be sure that any distant plume of smoke really is smoke and not a low-hanging grey cloud. Sight alone isn't enough and smell is unreliable – if the wind's in the wrong direction, the smoke will blow away from the beetles and never reach them.

Even when they've tracked down their goal – a site where a fire's just burnt itself out – the beetles still have problems. They need to land somewhere that's not too hot. The beetle is probably no better than the average insect at coping with heat; Schmitz reckons fire-beetle feet can only handle surfaces up to about 50 or 55°C. While too high for a safe landing, these temperatures aren't enough for smouldering wood and fields of hot ash to give off a glow during the day that the beetle's eyes could detect.

Hot-footing it

To succeed in their mission to reproduce, the beetles must find a fire and land somewhere that won't singe their feet. Both these challenges have something in common: heat. As anything warmer than absolute zero gives off infrared radiation, all a beetle need do to solve its problems is recognise infrared. 'It's very important that you have infrared sensors so you can detect where there's a hot spot or not,' says Schmitz. That stops you landing on the forest equivalent of a hotplate. It helps you find that hotplate too: when the fire beetle sees a plume on the horizon, its infrared detectors can confirm it's smoke from a fire and not a cloud.

We said 'all' a beetle need do. Detecting heat from a fire isn't easy, unless you're right up close. Imagine curling up on a sheepskin rug by a log fire on a cold winter's night, the golden flames flickering hypnotically as you bask in the warmth. It's perfect until you realise your drink is in the kitchen. Once you've moved back to the door, you hardly feel that warmth on your skin any more. The infrared radiation is still there but it's spread out to fill the whole space so it's less intense. If that's the drop-off over a few metres, what are your chances of feeling any heat from a fire way off on the horizon?

A fire beetle's infrared detection system must be better than human skin. All insects have heat receptors at the ends of their antennae, which – like the thermoreceptors in our skin – detect infrared only at high intensities. But fire beetles have a second, much more sophisticated, set of receptors, near the hips of their middle pair of legs, just under the wings. These detectors are good, but from how far away can they find a fire? To answer that question, Schmitz and his colleague Herbert Bousack of the Peter Grünberg Institute at the Forschungszentrum Jülich, Germany, turned to a blaze that burned at an oil depot in California in the 1920s. Their forensic detective work, published in 2012, was almost as astonishing as the beetles' heat-sensing power itself: the team consulted forest authorities, local newspapers and staff at Sequoia National Park in their quest to find out the fire's history.

What's so special about an old oil fire in the US, halfway round the world from Germany? What's wrong with modern fires? In woods? And surely the easiest way to work out the distance a fire beetle can spot a fire from is to slap a transmitter on its back and track its flight? But since these beetles are only a centimetre long, we haven't yet built a box of tricks small enough for them to carry far. The flight range of a heavily laden and tired fire beetle is likely to be much smaller than the distance it can spot a fire from. That's why Schmitz and Bousack had to think laterally. By turning to the archives, they learned more about a fire that raged for three days in August 1925 after a lightning strike and was visited by 'untold numbers' of fire beetles, despite being a long way from any forest.

The oil depot that caught fire lay several miles east of the town of Coalinga in California's Central Valley, where it's too dry for trees. The fire beetles must have come from the nearest forests, either around the San Benito Mountain 25km (15 miles) to the north-west or in the foothills of the Sierra Nevada mountains, 130km (80 miles) to the east. Ploughing back through the records revealed that the San Benito Mountain forests hadn't burned for a couple of years so they weren't a likely home for hordes of young fire beetles. Near the Sierra Nevada, on the other hand, there'd been major fires in each of the two years before the oil depot break-out. So the chances are that fire beetles flew to the oil fire all the way from the Sierra Nevada forests, a distance 13 million times the length of their bodies.

The heat, the heat

This historic detective work gave half the answer – a fire beetle can probably find a fire that's 130km (80 miles) away. It was time for some physics. Bousack turned to modern fire analysis – the 'pool-fire' kind that fire-protection officers use to determine how much heat a fire will throw at neighbouring buildings – to calculate the amount of infrared radiation beaming out from the Coalinga oil fire. 'The light of the fire was so great that one could easily read by it in town, a distance of nine miles,' the *Coalinga Daily Record* reported, adding that the flames shot hundreds of feet into the air.

Right next to the fire, which would have reached 1,000°C, Schmitz reckons, its irradiance – the amount of energy per square metre per second – was as high as 148,000W/m^2 (watts per square metre). Underfloor heating running nearly at full whack has an irradiance of just 100W/m^2. But 130km (80 miles) away in the Sierra Nevada forests, the infrared radiation, as in our fireside scenario, would be much less intense, with an irradiance some 100 million times smaller than at its source (0.13 milliwatts per square metre to be precise). Slight temperature fluctuations in the surroundings will mask an amount of heat this low, which makes detecting it quite a feat. Even man-made infrared detectors must be kept below freezing to be this good. Yet *Melanophila* beetles' heat detectors are so fine-tuned that, in theory, Schmitz says, they could be able to sense infrared radiation in its smallest possible form - a single photon.

'When we inspect freshly burnt areas, from time to time we find scorched insects – other beetles, spiders and flies – but we have never found a fire beetle that's been burnt,' says Schmitz. 'These animals can locate the hot spots during flight and prevent themselves from landing.' Which is just as well, because as soon as the fire beetles arrive at the burnt zone, they have mating to do. The ash-bestrewn remains of a smoking, black-stumped forest don't sound romantic, but for fire beetles they're perfect. The females beetle about, laying their fertilised eggs under the bark of the freshly burnt trees. When they hatch, the larvae chomp away on as much dead wood as they like, without fear of the tree fighting back. The following summer, the larvae turn into adult fire beetles. As a bonus, this year's adults are able to eat the bodies of other insects and small vertebrates killed by the fire. Schmitz has seen the Australian fire beetle 'feeding on a scorched lizard as a kind of barbeque'.

Ear's the answer

To detect such low levels of heat, the fire beetle almost 'hears' infrared, using hair-like mechanoreceptors lying inside small spheres filled with liquid. 'Mechanoreceptors are the most sensitive [receptors] in the animal kingdom,' explains Schmitz. 'They can sense movements in atomic dimensions, in the nanometre or even sub-nanometre range.' A nanometre is a millionth of a millimetre (go just a little smaller and you're inside an atom). We have mechanoreceptors too, in our inner ears, where sensory hair cells line fluid-filled chambers. Sound waves hitting the membrane of our eardrum travel through the three bones in our middle ear and on into the fluid in our inner ear, where they bend the hairs. The hairs convert the mechanical movement of this bending into an electrical signal in our nerves.

We've heard enough about ears for now; how do the fire beetle's infrared sensors work? Each detector is a pit about 0.3mm long, 0.12mm wide and 0.1mm deep. It's filled with between 70 and 90 tiny domes about 0.02mm across, clustered together like the lenses in an insect's compound eye. The domes are made of cuticle, the same material as the insect's exoskeleton. Inside, a teardrop of hard cuticle extends down into a sphere. Like the ad for a tempting brand of chocolates, the sphere is hard on the outside and spongy in the middle. The small spaces inside the sponge are full of liquid and the pressure-sensitive tip of a sensory cell is anchored at the core. When infrared radiation hits the detector, it warms the liquid, which expands and squeezes that tip, making the nerve cell fire an electrical signal that heads to the beetle's brain.

Sensitive type

That's all very well, but how does the beetle deal with the thermal fluctuations, or noise, masking the fire signal? This is where Schmitz gets really excited; he has a couple of theories for how the detector can be so sensitive. Either, he reckons, the beetle uses 'stochastic resonance', deliberately applying extra random noise that interferes with the signal hidden in the existing noise, boosting it enough for detection, or the beetle exploits 'active amplification'. In flight it flaps its wings 100 times a second. It could use this motion to vibrate the spheres inside its heat-detecting domes, priming them to pick up the slightest change in infrared levels. This may be why the detectors sit just beneath the beetle's wings. The amplification would mean that just one or two photons of infrared entering a sphere on one side of the insect are enough to make its nerve respond, showing the beetle which way to fly to find heat. It's the same approach as the cochlear amplifier system in humans and other mammals, which vibrate the hairs in their inner ears to pick up quiet sounds. Recently scientists discovered active amplification in insects too; *Drosophila* fruit flies use it in the 'ears' in their antennae. Back in Germany, Schmitz is trying to prove that the fire beetle uses active amplification to 'hear' infrared.

In some ways Germany is not the right location for Schmitz. Conforming to the national stereotype, the country's firefighters are so efficient that fire beetles are now extinct; there's no longer enough burnt-out forest. So the beetles are no fan of the firefighter. In turn, firefighters in other countries aren't too keen on the beetles. As these emergency workers hose down the last few hot spots after a forest blaze, the beetles sometimes get inside their suits and bite. Something similar happened at our 1943 football game. We don't know what the spectators said when the beetles attacked, but this incident definitely happened. Attracted by around 20,000 lit cigarettes, huge swarms of *Melanophila consputa* and *Melanophila acuminata* descended on the game and bit fans, as reported in the Entomological Society of America's *Journal of Economic Entomology* in 1943. If fire beetles must now live alongside humans, are their heat detectors too sensitive for their own good? As well as turning up at football games, these insects have swarmed at cement kilns, smelters, tar producers and round the hot vats at sugar refineries. Close but no cigar.

Heat in short

Time for a summary. The garter snake, dog and mosquito exploit the physics of heat by huddling up, shaking their floppy skin, or excreting blood to keep themselves at the temperature where their bodies work best. The Japanese honeybee, California ground squirrel and fire beetles take this philosophy one step further, using heat to defend themselves or their young, whether from hornets, rattlesnakes or trees. The rattlesnake 'sees' heat, while the fire beetle 'hears' heat to learn more about its surroundings.

We've seen how heat is a form of energy transfer and temperature is a measure of how much an object's atoms or molecules are jigging about, which also shows its ability to transfer heat. That heat transfer can take place by conduction (such as when garter snakes snuggle close), convection (as in bathwater or pockets of air trapped in fur) and radiation (like the Sun beaming energy across space as electromagnetic waves). The fire beetle detects such radiation by sensing the force when a heated liquid expands and pushes on a hair-like receptor. There's more on forces coming up next, including a dragon with a surprisingly delicate feature, lizards with a head for heights, a shrimp with beefed-up elbows, and death-defying mosquitoes.

CHAPTER TWO

Forces: The Big Push

A LIGHT-HEADED DRAGON * RAINDROP-DODGING
MOSQUITOES * A SHRIMP THAT PUNCHES ABOVE ITS WEIGHT
* THE WORLD'S FASTEST ANIMALS * A SNAPPY ANT
* THE LIZARD THAT THINKS IT'S SPIDERMAN

Here be dragons

As any cat lover knows, when a moggy rolls onto its back, front paws flopping, and seems to invite you to tickle its tummy, it doesn't necessarily mean it. One moment you're stroking soft fur while its owner purrs; the next there are 18 claws snagging painfully into your skin as you snatch your hand away. If you're unlucky, the cat will use its teeth on you too, clamping down its long sharp canines to puncture your skin.

Imagine how much worse you'd feel if the animal wasn't a 6kg (13lb) cat but an 80kg (175lb) Komodo dragon (*Varanus komodoensis*) lumbering its way towards you through a jungle clearing, head swinging from side to side as it 'tastes' your smell in the air with its forked tongue. For most of us this is an unlikely scenario but for people on the Indonesian islands of Komodo, Rinca, Flores, Gili Motang or Gili Dasami, it's a real threat. Although the endangered reptiles are wary of humans, sometimes they go rogue and attack. They'll even turn cannibal and eat their own juveniles if they can catch one that isn't hiding up a tree.

So the Komodo dragon is the stuff of nightmares. This monster or, as biologists prefer to call it, monitor lizard, lives in savannah and low-lying tropical forest. At roughly 2.5m (8ft) long, the animal is the world's largest living lizard and built like a low-slung Godzilla. With a stocky body covered in armoured scales, a long tail, stumpy legs, and its 'elbows' pushed out to the side so that it walks as if doing press-ups, this dragon lives up to its name. It can even stand on two legs. As the *Hitchhiker's Guide to the Galaxy* author Douglas Adams wrote in his 1990 book *Last Chance to See*, co-authored with naturalist Mark Carwardine, the reptile could be behind the 'Here be dragons' warning that explorers of old wrote on maps showing land they didn't like the look of. Perhaps the flicker of a yellow tongue confused early explorers into assuming the reptile was breathing fire. At the very least the Komodo dragon has powerful halitosis – its mouth is a haven for bacteria. Let's look at which is worse, the Komodo dragon's breath or its bite.

When Stephen Wroe of the University of New South Wales, Australia, investigated the jaw strength of the Komodo dragon, he and his team found something peculiar. Before we discover what, we should add that Wroe's real interest wasn't in the Komodo dragon but in a reptile twice the size: megalania (*Varanus priscus*), a 5m (16ft) monitor lizard that roamed Australia until roughly 100,000 years ago. The Komodo dragon is this giant's most comparable relative alive today. By studying these surviving reptiles, Wroe, who's now at Australia's University of New England, hoped to discover more about megalania's meals of choice.

To research the Komodo's bite without losing an arm, Wroe, Domenic D'Amore and colleagues tempted 10 Komodo dragons from four zoos around the world to clamp their jaws onto a piece of meat attached to two bars of aluminium. These weren't just any metal bars. They were coated with rubber to stop the animals damaging their teeth, and rigged up with strain gauges that recorded a voltage as the bars flexed under the load. Strain – the amount that something deforms as a proportion of its length – depends both on that item's stiffness and on the stress, the load or force divided by the area it's applied over. Later the team translated the voltages they'd recorded into bite forces, using measurements of known loads they'd made beforehand.

The strongest-jawed Komodo dragon could bite with a force of 149 newtons (N or kgm/s^2, more on this unit later). And the others were less powerful. But a pet cat some 20 times lighter than the Komodo dragon has a 58N bite, so this 'interesting, sexy, iconic' reptile isn't punching above its weight. Humans, who weigh roughly the same as a Komodo dragon, have been known to bite with a force of 294N. On this scale, being chewed by a Komodo dragon is less bad than a gnawing from a football or rugby player. If its bite strength was all it had going for it, a jungle stand-off with a Komodo dragon wouldn't be that terrifying.

Light bite

Despite its feeble jaw snap, the Komodo dragon is one of only two living varanid lizard species to kill animals larger than itself. The reptile's favourite snack is carrion – meat that's already dead. When it does attack live prey it goes for deer, pigs, birds, invertebrates (animals without backbones, they're literally spineless), the occasional human, its own young or even water buffalo weighing a few hundred kilograms.

How does the Komodo dragon kill animals this big with a bite not much stronger than a cat's? It would be a surprise if your pet turned up at the cat flap with a freshly slaughtered pig in its jaws. According to Wroe, when going in for a kill, the Komodo dragon uses a 'can-opener' action. It moves its head to the side, bites down on its prey's throat or underbelly, then rotates its head back to the centre, deploying its 60 teeth in turn to slice open a curved wound. At the same time as biting, the reptile pulls back by moving its head down and 'back-pedalling' its legs towards its tail. This enables its strong neck and body to compensate for its feeble bite force. Since Komodo dragon teeth are up to 2.5cm (1in) long, the resulting gash in the victim's flesh lets loose a lot of blood.

'What you get is a ripping or tearing motion,' says Wroe. 'Given that the animal has very sharp serrated teeth it can cause a lot of damage. It's an advantage to the Komodo in that it doesn't need particularly big jaw muscles or a heavy solid skull; the downside is that it's not very well adapted to anything else.'

So the advantage of the Komodo dragon over the cat, when it comes to killing, is a 'can-opener' technique and a powerful body. To find out how much a strong body makes up for a feeble bite, Wroe and colleagues measured the pull force the animals exerted with the muscles of their head and neck on a piece of pork attached to a digital force gauge. The largest force applied, when the pork was just above the ground and the force gauge was 1.5m (5ft) higher, was 337N, much stronger than the bite force, where the record-breaker was just under 150N. The Komodo dragon's snap, twist, pull and slice technique tears animal bodies into chunks so that it doesn't need a strong jaw to crunch up bones, unlike the striped hyena (*Hyaena hyaena*), which has a bite force of 545N.

Monster force

Finding the Komodo dragon's killing ability superior to that of a small domestic cat isn't a fair contest. How does the reptile's attack strategy rank against those of animals its own size? On Komodo island, the Komodo dragon is often dubbed the 'land crocodile'. That similarity has struck Wroe too – he reckons the most obvious comparison for the Komodo dragon is another reptile, Australia's saltwater crocodile.

'Crocodiles have a much more solid skull and more massive jaw-closing muscles,' he says. 'A crocodile when it's chasing relatively big prey will take a bite into it, roll and twist. [The croc] can generate and withstand forces from a wide range of directions so it's a bit more haphazard.' The Australian reptile is more of a thug, roughly spinning its hapless victim around. The delicate-skulled Komodo bites, twists its head round and pulls in just one direction: back and down.

The Komodo dragon also has a different approach to big cats and other mammal meat-eaters, which use their powerful jaws and strong bite force to squeeze their victim's windpipe until the animal suffocates. Lions have a bite force of around 1,768N, much higher than a Komodo dragon. But their killing technique leaves the lions vulnerable to injury from thrashing hooves. 'If you want to clamp the neck down on a Cape buffalo you need a very strong set of jaws, and more than a little courage,' says Wroe. 'It's a risky game and lions often get hurt.'

The Komodo dragon is more akin to the lion in the *Wizard of Oz*, who has lost his bravery. The reptile is a cautious beast that reduces its risk of getting hurt. 'It can run in and quickly produce a fatal wound then stand back, rather than having to hang on for dear life whilst the animal slowly suffocates,' says Wroe. 'In the case of a Cape buffalo that could easily take 10 minutes or more. You've got to hang on and hope that its big brother doesn't come along and kick your head in.'

The now-extinct sabre-toothed cat *Smilodon fatalis* used tactics along the lines of the Komodo dragon's, Wroe believes. These creatures also compensated for their weak jaw-closing muscles with other body parts – this time with strong muscles for pushing their head downwards. 'They were adapted to wrestle and restrain relatively large animals using very powerful post-cranial [behind the skull, i.e. neck and body] muscles, and then use their head-depressing neck muscles to sink their canine teeth into the neck of large herbivores,' he says. 'It's a broadly similar strategy in that the animals kill prey by inducing massive trauma through blood loss.'

Poisonous smile

As if causing fatal bleeding wasn't enough, the Komodo dragon has another trick that lets it bite then hang back somewhere safe. Before 2006, biologists thought that bacteria in the dragon's mouth infected its prey with septicaemia. But if that were true, the inflammation would take days to kill so the predator would 'likely not enjoy the fruits of its labour'. Wroe's colleague Bryan Fry from the University of Queensland discovered that the Komodo injects its victims with venom, using a poison that contains shock-inducing neurotoxins as well as an anticoagulant to prevent blood flowing out of the bite wound from clotting. This blood loss/poison double whammy helps the Komodo dragon kill animals up to the size of a water buffalo, some four times larger than itself, by inflicting life-ending injuries swiftly, then retreating and waiting for its victim to die. There is definitely fire in this dragon's bite.

While the Komodo dragon has worked around its weak bite to come up with a successful – and unique – twist-and-pull killing strategy, the reptile's feeble jaw does cramp its style. If its jaw and skull were more robust, like the crocodile's, the animal could wrestle its prey, rolling around and exerting forces in all directions rather than pulling only sideways and backwards. So why doesn't the Komodo dragon have a stronger bite?

To discover the reason, Wroe, Karen Moreno and colleagues analysed the structure of a Komodo dragon skull using a technique known as finite element modelling. Engineers use this method to design anything that will carry a load, from bridges to car engine components and from aircraft to buildings. It splits up the item of interest into shapes such as triangles or squares, and solves the complex equations that describe how much each shape will deform under a stress. Because the shapes are joined together, the findings for each one feed into the calculations for the shapes next to it.

The results aren't perfect but they're good enough; it's easier to solve equations over lots of small shapes and combine the results than to handle everything at once. It's like dividing a spot-the-difference picture into a grid. You can check each small square one by one, rather than gazing at the whole image in confusion and despair.

Wroe and Moreno's model mimicked a Komodo dragon skull held in a museum, from a young adult male that was 1.6m (5.2ft) long, splitting it into 1.2 million 'bricks'. By imaging the skull in 3D with a computed tomography (CT) X-ray scanner at a hospital, the team assigned each brick one of four bone densities. Denser bone is stronger and stiffer so it deforms less when you apply a force. Wroe also looked at the structure of the jaw muscles to calculate forces and areas. The finite element modelling revealed that the Komodo dragon skull is bulky only where it's useful. The skull has thick regions of bone towards the back and in other areas where it's needed for support, and empty space elsewhere.

'The Komodo dragon's skull is highly optimised for very specific functions,' says Wroe. 'That means it can save a lot of weight by using less bone and less muscle. Its skull is basically what we call a space frame structure – it works on beams or struts.'

The skull resists large forces in the sideways and backwards directions (towards the Komodo dragon's tail) at the same time, Wroe found, making it ideal for the way that the animal boosts its weedy bite with a ripping twist and a strong pull. Its bone structure is so fit-for-purpose that the skull experiences lower stress when it undergoes a pull force together with a bite force, rather than a bite force alone. Although this weight-saving design limits the Komodo dragon's jaw strength and prey-wrestling options, the animal needs less energy to support its head and move around. And that means it must find less food.

Reptiles lost

What about Wroe's true interest, the long-dead megalania? His force-analysis detective work shows that if this extinct giant attacked its prey in the same way as its modern Komodo relatives, it could have eaten animals like the giant wombat, a marsupial that grew as big as a large rhino. The megalania is long gone, but what does the future hold for its Komodo dragon cousin? While the reptile may not sound vulnerable, killing the occasional human as it does, 'vulnerable to extinction' is its official ranking according to the International Union for Conservation of Nature. There are just 4,000–5,000 of the animals left, perhaps as few as 350 of them breeding females, and their habitat is under increasing threat. But hey, at least this ranking saw the Komodo dragon join the aye-aye, pink river dolphin, kakapo and others in *Last Chance to See*. It also received a follow-up visit from Stephen Fry with Adams's co-author

Mark Carwardine for a BBC TV documentary aired in 2009. When it comes to destruction, the Komodo dragon's breath is not as bad as its bite, but humans may ultimately prove to be more destructive than both of them.

A weighty problem

It's time to link the Komodo dragon to the physicist who sewed up the laws governing forces and motion. While we're not sure if Sir Isaac Newton (1642–1727) had bad breath or went around biting people, Google reveals that sci-fi writer Robert Heinlein named a dragon in his 1951 *Between Planets* after this British physics genius. The creature was from Venus, not Komodo, but it's fitting that Newton is in a story about other planets – among many great achievements, he was the first to understand why planets orbit the Sun in ellipses, not circles.

Newton was an unusual character. As a schoolboy he amazed his classmates by constructing a model windmill powered by mice running on a treadmill, and caused an early UFO scare by attaching lanterns to kites. Later he became the first science celebrity and the subject of several poems, including one by William Wordsworth. From 1978 until 1988 Newton featured on the Bank of England's one-pound note, the curls of his wig falling down his shoulders alongside his cravat as he sits, book in lap, to the side of a telescope. And the unit of force is named in his honour: one newton (or 1N) is the force needed to accelerate a mass of 1 kilogram at a rate of 1 metre per second each second, or 1kgm/s^2.

Some say that, rather than kites or mouse power, it was the Moon that was on Newton's mind when he saw an apple fall from a Lincolnshire tree in 1666 or thereabouts. He'd had to leave Cambridge University when the plague struck town, retreating to the orchards at his family home. Crucially, he realised that the Moon and the apple were linked: the gravitational force causing the apple to fall to the ground is the same as the force keeping the Moon in its orbit around the Earth and preventing it from travelling in a straight line. And that force drops off with the square of the distance from the centre of the Earth; this inverse-square law is partly responsible for the elliptical orbit of planets round the Sun.

Newton wasn't the first to think about gravity. Italian astronomer Galileo Galilei (1564–1642) had allegedly chucked balls made of different materials off the Leaning Tower of Pisa to prove they'd take the same amount of time to reach the ground as gravity accelerated their speed from zero (as long as the air resistance was low). But this is probably just a legend, as is the tale of the apple falling onto Newton's head while he slept under a tree. Newton was walking in the garden thinking, not lazing about, historians believe, and saw an apple fall rather than receiving a direct hit. But that's not as good a story. William Stukeley (1687–1765), a clergyman and druid who pioneered investigation of the ancient British site Stonehenge, noted Newton's own words in his *Memoirs of Sir Isaac Newton's Life* (1752) following a chat with the great physicist beneath some apple trees in London in 1726:

> *Why should that apple always descend perpendicularly to the ground, thought he to himself. Why should it not go sideways, or upwards, but constantly to the earth's center? Assuredly, the reason is, that the earth draws it. There must be a drawing power in matter, therefore does this apple fall perpendicularly, or toward the center. If matter thus draws matter, it must be in proportion of its quantity. Therefore, the apple draws the earth, as well as the earth draws the apple.*

Both the apple and the Earth exert a force on each other. We don't notice the force of the apple because the Earth is so large that the apple moves it by an amount too small for us to detect. There's more on the power of very small forces later when we discover how the gecko walks on the ceiling. Newton encapsulated how objects exert forces on each other in his third law of motion, which he wrote down, together with his two other laws of motion and the universal law of gravitation, in his 1687 classic *Principia Mathematica*.

If you're keen, you can read this in its original Latin on the University of Cambridge digital library website. But bear in mind that when Newton published, one aristocrat offered £500 to anyone who could explain what the *Principia* meant. For those who'd prefer a summary in modern-day English, Newton's third law of motion states that for every action there is an equal and opposite reaction. In other words, there's a 'pushback' for any force acting on an object from a counterpart force exerted in the opposite direction.

As well as his physics nous, Newton was famous for his mathematical genius, so we shouldn't write about his laws in a non-logical sequence. Having started with number three, we'd better work backwards in order. In modern English, Newton's second law states that the change of momentum of a body is proportional to the impulse impressed on the body, and happens along the straight line on which that impulse is impressed. When TV sports commentators say a team 'has momentum' they mean it's 'on the move' or 'hard to stop'. That definition is reasonably close to the official physics version: momentum is an object's mass multiplied by its velocity (its speed in a specific direction). Impulse, meanwhile, is the force multiplied by the time it's applied for. So Newton's second law also means that an applied force on an object equals the rate of change of its momentum with time. As the mantis shrimp and trap-jaw ant will show us, there's another way to think about it too. But first, and before we get to law number one, let's find out how the mosquito keeps Newton's second law for a rainy day.

Dodging raindrops

Can mosquitoes fly in the rain? It's one of those questions
it doesn't even seem worth asking. Like, was Albert Einstein
a genius? Mosquitoes thrive in the tropics, where it rains a
lot. They need pools of water to drink from and to lay their
eggs in. So of course they can fly in the rain. Who's seen
mosquitoes dashing for cover in a storm?

But hang on a minute. Raindrops plummet from the sky
at nearly 10m (33ft) per second, or more than 35km per
hour (22mph). The heaviest weigh 100mg, about 50 times
the mass of a mosquito. Each drop is roughly the same size
as the insect, but mass for mass, a stationary mosquito
receiving a smack from a raindrop is like a 5,000kg (5-ton)
lorry cannoning into a 100kg (220lb) human. In a downpour,
a mosquito is hit by a raindrop roughly once every 25
seconds. Flying in the rain sounds like an online road-
crossing game where the mosquito must constantly dodge
impacts with heavy vehicles. How does it survive?

This question intrigued Andrew Dickerson, who we met
in Chapter 1 studying how wet dogs shake themselves dry.
Based at the Georgia Institute of Technology, Dickerson had
a ready supply of mosquitoes from the nearby Centers for
Disease Control and Prevention in Atlanta. How, he asked,
do these fragile insects cope with strikes from fast-moving
raindrops? It was more than idle curiosity: mosquitoes spread
the parasites that cause malaria, so knowing more about
their flight could help prevent the disease.

The work wasn't easy. Dickerson's early experiments
were 'unmanageable' and a 'complete failure'. All he ended
up with was a wet floor and 'lots of frustration'. (With all
those soaking dogs, you'd think he'd be used to damp
floors ...) Dickerson got there eventually, gaining a PhD for
his efforts. Apart from giving us a new-found respect
for mosquitoes, which survive most raindrop collisions,
Dickerson's research showcases Newton's second law of
motion and the concept of momentum.

First let's think about that 5,000kg truck we mentioned. Imagine it's thundering towards you at 10 metres per second. It has a momentum of 50,000kgm/s. That's a lot. You, ambling along, have much less momentum as you're lighter and slower. If the truck collides with you – sorry this is gruesome – it'll fling you backwards at high velocity. You'll suffer a big change in momentum because your speed has gone up so much and also, if you were walking towards the truck, your movement has changed direction. The force you'll feel is your change in momentum divided by the length of time you were in contact with the truck, since the applied force on an object, under Newton's second law, equals its rate of change of momentum with time. As the collision was over so quickly, you'll feel a huge force. Ouch. After flying through the air, you'll smash into the ground and your speed will drop to zero. Again, you'll experience a swift change in momentum and a big force. It's why traffic collisions are so nasty.

Never rains but it pours

What does this mean for our rain-drenched mosquito facing its own equivalent of a truck smash every 25 seconds? To find out, Dickerson built a small plastic cage, about 5cm (2in) wide and 20cm (8in) high, for *Anopheles gambiae* and *Anopheles freeborni* mosquitoes. On top of the cage he put a water-soaked mesh with holes far smaller than the insects so they couldn't fly out. After leaving the cage on the floor of the atrium at his lab, Dickerson fired drops of water at the mosquitoes from a height of 2 metres (6.5ft). The idea was that when a drop hit the watery mesh, it would eject a new drop into the cage. These tests went well but the droplets didn't hit the mosquitoes as fast as raindrops. So Dickerson replaced the meshed lid on his mozzie cage with a cover containing a penny-sized hole, and climbed to a 10-metre-high (33ft) balcony to release water. Falling from a greater height, these drops should reach the speed of rain. With 4,000-frames-per-second video cameras at the ready, Dickerson was all set to record the collisions.

But getting the drops into the cage proved impossible, even though some of the mosquitoes managed to get out, leaving Dickerson with no results other than a sodden floor. His solution was to let pressure, not gravity, accelerate the droplets by firing a jet of water through a spray nozzle fixed at the top of the enclosure. This time his experiment worked, and Dickerson shot drops past the mosquitoes at 10 metres (33ft) per second, the rate of naturally falling rain. Analysing his footage in slow motion, Dickerson noticed that the insects made no attempt to avoid impact; he caught six mosquito–raindrop collisions on camera. When a raindrop strikes a mosquito, he discovered, it usually hits the animals slightly off-centre – on their legs or wings. The drop knocks the insect off course but it recovers fast, sometimes within a hundredth of a second.

The worst type of collision is when a hovering mosquito is whacked directly between its wings. The droplet and mosquito tumble downwards together, entangled like lovers, with the impact speeding the insect from zero to about 2.1 metres per second in just 1.5 milliseconds. That gives it an acceleration of roughly 1,400 metres per second per second (2.1 metres per second divided by 1.5 milliseconds). That's more than 140 times the acceleration due to gravity. Ride on a roller coaster, or in a Formula One car or a space rocket, and you'll probably cope with about five times the acceleration due to gravity. Anything higher and you'll faint, feel your eyeballs pop out, or die, although Dickerson says one US doctor survived accelerations of 25 times gravity in military flight tests during the Cold War. Fleas, however, which are incredible jumpers, accelerate at about 135 times the rate of gravity. 'It is amazing what levels of acceleration insects can survive,' says Dickerson.

Since mosquitoes are so light, their momentum change is relatively small. According to Newton's second law, assuming the insect weighs 2mg, the collision force it suffers from a raindrop is 0.003N (its rate of change of momentum: 2mg multiplied by 2.1 metres per second, divided by 1.5 milliseconds). That's no big deal for a mosquito. Separate unpleasant-sounding 'compression tests' that Dickerson carried out showed that mosquitoes – thanks to their armour-like outer skin, or exoskeleton – can survive a squashing by forces 10 times higher than those from a raindrop.

Another key to feeling less force in a collision, even though it sounds like the opposite of a good plan, is to make the impact last as long as possible. Mosquitoes achieve this by sticking to, and travelling along with, the water droplet. Boxers use this 'longer impact' principle if they realise they're about to take a blow. They relax their neck muscles and let their head move backwards, 'riding the punch'. Their opponent's glove makes contact with the boxer's head for longer, lowering the force of impact. The same idea comes into play if your car's dashboard air bag inflates – the impact lasts longer because the air bag slows your motion more gradually than the harder surface of the dashboard itself, reducing the collision force.

Raindrops keep falling on my head

So the insect can survive the initial raindrop hit. What happens next? Once they're together, the raindrop and mosquito continue their attachment. Dickerson saw one pair travelling for about 39 millimetres (1.5in) – roughly 13 mosquito body lengths. The drop's not stuck to the mosquito, which is covered in hairs that repel water. It's more of a gentle embrace. Eventually, though, the mosquito slides away from the raindrop and lands on the side of the cage. After a short rest, the insect flies off, unharmed. Each flying mosquito hit by a drop lives on. The drop isn't damaged either: it stays intact while the mosquito hitches a ride. In his early tests with drops falling more slowly,

Dickerson took higher-resolution video that showed that the drops buckle and deform, especially if they're small. But surface tension – a phenomenon we'll return to in Chapter 3 – stops them falling apart.

Drop dead gorgeous

Mosquitoes can't, however, relax in the rain. Until now, we've assumed the insects have as much space to fly in as they like. But if they're too low when hit by a water drop, the insects are in danger of being propelled into the ground, where their momentum will change big-style. Or if the drop lands on a mosquito sitting on a tree branch, it's also in for trouble. A raindrop hitting a mosquito in mid-air goes along for the ride, but when it splats into a mosquito on the ground or a branch, the drop breaks up. If all the momentum of the raindrop transfers to the mosquito, the insect gets slammed with a force equal to the momentum of the raindrop – its mass (0.1g) multiplied by its velocity (10 metres per second) – divided by the time it takes the drop to spread out (around 2 milliseconds). That's a force of about 0.5N, roughly 50,000 times the weight of a mosquito. It's enough to kill the insect. Bye.

The key to a flying mosquito surviving a raindrop hit lies in its low mass and its ability to 'ride the punch'. That's cracking news for mosquitoes, though *not* so great if you don't like these insects. As we mentioned in Chapter 1, mosquitoes transmit deadly diseases to humans, so keeping the pests at bay is vital in many areas of the world. Insecticide sprays are our weapon of choice. Squirting these poison-laced liquids into the air creates a fine mist of mosquito-killing droplets, each about one-thousandth the size of a raindrop. Yet insecticides aren't great for the environment: who wants to deal with toxic chemicals? More recently, Dickerson has looked at how *Anopheles freeborni* fly in dense fogs of water droplets, and a variety of gases, with no toxins in sight. Exposing the mosquitoes to gases that are twice the density of air makes them fly erratically, pitching and rolling before tumbling to the ground, he found. The researcher reckoned the gases create unusual drag forces on the insect and disrupt its halteres, the body parts that sense its position. And that gave him an idea for an alternative to toxic sprays. As water droplets don't stick to the waxy, water-hating hairs that cover mosquitoes, Dickerson proposes sprays made entirely from soybean oil, which will cling to the insects better. This natural material, already used as a base for chemical insecticides, would stick to the halteres and upset the insects' flight, keeping them away from the person armed with the spray. All that's a long way off. For now, if you want to stay safe in malaria zones, Dickerson advises conventional creams and ointments. 'Even in the rain,' he says, 'wear your insect repellent'.

Join the club

Mosquitoes avoid death by raindrop by employing Newton's second law: 'the change of momentum of a body is proportional to the impulse impressed on the body ...' And that's the same as saying that an applied force on an object equals the rate of change of its momentum with time. As long as the object isn't getting heavier or lighter as it moves, Newton's second law also means that the applied force equals the mass of the object multiplied by its rate of change of velocity, that is, its acceleration. In symbol form this is $F=ma$, an equation that in 2004 was voted third favourite equation of all time by readers of *Physics World* magazine. It could well be the favourite equation of the harlequin mantis shrimp too, as we'll find out after talking about food again. Not water buffalo or apples this time, but crab, mmm. . .

Dining out on crab is a messy experience, involving metal tongs, plastic aprons and wayward bits of flesh flying across the restaurant. Another lover of crab meat, the harlequin mantis shrimp (*Odontodactylus scyllarus*) doesn't have such tools, living, as it does, in the tropical waters of South-east Asia and the South Pacific. Despite its name, this brightly coloured 10cm (4in) crustacean is no clown when it comes to smashing up crabs for the juicy meat inside. Also known as the peacock mantis shrimp, *Odontodactylus scyllarus* has developed its own tools for cracking the shells of crabs, snails and bivalves such as mussels: it has extra-strong 'dactyl' segments on the end of its second pair of appendages that it uses as clubs. 'Dactylus' means finger in Greek, although this shrimp's dactyl is more of a bulbous elbow that it thrusts forward from the equivalent of its upper arm. To a poet, a dactyl is one accented syllable followed by two unaccented ones, like the phrase 'mantis shrimp'.

As if flagging the danger they represent to a shelled creature, the harlequin mantis shrimp's dactyls are bright orange. Much of the rest of the shrimp is brightly coloured too – it has round purple eyes goggling above its turquoise face, a white and red spotted front, a green body and 10 red legs. The animal would suit a Disney cartoon (the less flamboyant Jacques of *Finding Nemo* fame is a banded cleaner shrimp). Shaped like a giant, bendy-backed woodlouse, the shrimp not only looks stunning but when it creeps out from its U-shaped tunnel in the sea floor near a coral reef, it can deal a stunning blow.

'Mantis evolved a few hundred million years ago,' says David Kisailus of the University of California, Riverside, who describes the shrimp as being like a heavily armoured caterpillar. 'Originally it was a spear fisherman. Its dactyls were barbed and it would use this to spear fish. As other things changed, it "decided" to eat crabs and other animals with shells.'

But crab shells are strong and tough. How can a 10cm shrimp exert enough force to overcome the crab's defences? 'It uses the "elbow" region of its dactyl to try and break open these harder elements,' says Kisailus. 'Over time the mantis with well-endowed elbows split off into a different group. So there are the spears and the smashing ones.'

Well-endowed these elbows may be, but size isn't everything when it comes to crab shells. Fortunately, this shrimp is no wimp. It can exert peak forces on its crab dinner of more than 700N, over a thousand times its own weight (weight is the force on an object due to gravity – the weight of a shrimp with a mass of 60g on Earth is 0.6N, that is $0.06kg \times 9.81m/s^2$, the acceleration due to gravity). Scientists have tested this by putting food paste on a force-measuring device; the mantis shrimp bashed the 'load cell' repeatedly as if it was a crab or snail. 'It has to generate these kind of forces to supply enough energy to form cracks in the shells of organisms it's preying on,' says Kisailus.

The crustacean is a physics geek; it powers its punch through Newton's second law (F=ma). To make a large force, the shrimp must have heavy clubs or accelerate them fast. There's a limit to how massive its dactyls can be – the shrimp doesn't want to be weighed down by heavy mouthparts as it scuttles around the sea floor, and it can't make its clubs heavier the second it spots something to eat. But it can accelerate them. And the larger this acceleration the better. The shrimp accelerates its dactyls at up to 100,000 metres per second per second (m/s^2), a rate more than 10,000 times that of gravity, and one of the fastest accelerations in nature. The clubs reach speeds of more than 80km per hour (50mph) before impact. In video footage of a mantis shrimp bashing a bivalve, the club slams out almost too fast to see. But the damage to the shell is all too visible – after a few resounding blows, there's a caved-in hole. When combined with another physics phenomenon sparked because the tip of the club is moving so fast, the shrimp can apply forces of around 2,500 times its body weight. More on this later. Suffice it to say that a crab in this scenario will crack. Being hit like this would make anyone crabby.

The need for speed

How does the harlequin mantis shrimp accelerate its club so fast? It can zoom its claw from a speed of zero to over 80km per hour (50mph) in the blink of an eye. Or, given that it can take humans 0.3 seconds to blink and the mantis shrimp needs just 3 milliseconds (0.003 seconds) to strike, 100 times quicker than the blink of an eye. The feat is all the more impressive since animal muscles don't spring into action that fast. Speedy animals like the cheetah and peregrine falcon accelerate relatively gently, starting slowly and taking their time to build up speed over a considerable distance. Touted as the fastest land animal, the cheetah accelerates at about 9 m/s^2 (from 0 to 96km per hour, or 0 to 60mph, in 3 seconds) and can reach more than 100km per hour (62mph). The fastest moving animal of all, the peregrine falcon, uses gravity to swoop towards the Earth at more than 300km per hour (185mph). So how does the mantis shrimp accelerate its club faster than its own muscles can manage? The canny shrimp employs a technique known as power amplification, as Sheila Patek of Duke University, US, explains. It's nothing to do with speakers at a rock gig, more like a weapon developed in ancient China.

'If you were going to throw an arrow with your arm to kill a deer, it wouldn't do much good,' says Patek, 'but you can use those exact same arm muscles on a crossbow to store elastic energy in advance and release that energy with a latch. The crossbow amplifies the power – the rate at which the energy is released.' This amplification accelerates the arrow much faster than you could by hurling it. 'There's this crazy trade-off due to the way muscles have evolved,' says Patek. 'Either a muscle can contract quickly without much force or it can contract slowly but with a lot of force. You can't have both, meaning that you can't get a lot of power out of your arm muscles without the help of something like a bow.'

A crossbow stores energy by ratcheting up the tension in a cord that your arm muscles pull backwards while the bow keeps it taut. The harlequin mantis shrimp stores the energy from its muscles in its own stiff exoskeleton, which it compresses.

'You're probably familiar with antagonistic muscles in your arms,' says Patek. 'You have one set of muscles that bends the forearm at the elbow and another set that extends it.' The mantis shrimp has a whopping set of 'closing' antagonistic muscles. At the same time as it acts to open its strongly muscled claw, the shrimp contracts its closer muscle. It may be relatively weedy but this closer muscle has a specially modified tendon that works like a latch. 'The closer muscle latches the appendage closed while this really massive opener muscle compresses a spring [part of the shrimp's exoskeleton],' says Patek. 'The closer muscle is a fast-relaxing muscle. When it turns off, this releases the latch and allows the appendage to swing out over a very short time period.'

The exoskeleton stores the energy from the shrimp's forceful but slow-acting antagonistic muscles trying to open its claw, before its fast-relaxing closer muscles release the latch, unleashing all the slowly built-up energy in a short sharp burst. By making everything happen more quickly than it can through un-power-amplified muscle power alone, the shrimp boosts the acceleration of its dactyl to levels that are near record-breaking for the natural world.

Blowing bubbles

Although in the speed (rather than the acceleration) race, the harlequin mantis shrimp isn't an outright winner against the cheetah and peregrine, it certainly wouldn't feel like a loser to a crab or mollusc under attack. As if being hit extremely fast by a powerful claw wasn't bad enough, there's worse to follow, just half a millisecond later. The shrimp uses another type of physics to beef up its blows.

According to Patek, a dactyl moves so fast through the water that it creates two regions – one flowing very fast and the other, right next to the dactyl, flowing very slowly. That means the Bernoulli effect, named after Swiss mathematician and physicist Daniel Bernoulli (1700–82), kicks in. The intersection between high and low flow creates a patch of low pressure that makes the water molecules separate from each other into a vapour bubble, in a process known as cavitation.

A bash from a bubble doesn't sound too bad, but this is no soap bubble bursting with a gentle pop. On hitting the snail shell, the cavitation bubble implodes in a ball of heat, light and sound. 'It emits an enormous amount of energy, [reaching] 7,000°C like the surface of the Sun,' says Patek. You can see an example of the flashes in a video on Patek's web page – it looks as if the shell is being zapped with a series of sparkles from the end of a magician's wand. In a spooky coincidence, power amplification is not only the reason why the shrimp can accelerate its claw so fast but also why the resulting cavitation bubbles do so much damage. 'Lots of energy is required to separate those water molecules,' Patek says, 'and if you shorten the amount of time for them to get back together you get a huge amount of energy release over a short amount of time.'

Just when the snail has recovered from the first blow and then its cavitation bubble, the shrimp's other claw hits, followed by more cavitation. The snail gets a quadruple whammy of impacts, like being hit by a jackhammer or a boxer who sets off a firework in your face after every punch, adding insult to injury. Punch, boom, punch, boom. Small consolation that the process is over in less than 0.8 milliseconds. The forces peak at 1,500N, more than 2,500 times the shrimp's body weight – the cavitation bubble is even more devastating than the club's strike force. For a human roughly the same mass as one of the authors, that would be equivalent to hitting with a force of 1.5 million N. A boxer's punch, in contrast, lands with up to 5,000N of force. A punch impacts for longer than a mantis shrimp club, increasing its impulse, so these values don't provide the full picture. Still, note to self: do not mess with the mantis shrimp.

Cavitation is such a powerful weapon that the pistol shrimp, another crustacean, shoots out cavitation bubbles to stun its prey. 'Compared to smashing mantis shrimp they're really loud,' says Patek. 'They pop continuously because they're shooting their cavitation bubbles at each other and their prey. There's no avoiding them – you can hear them outside the tank, you can hear them in the hallway popping away. The sounds of the pistol shrimp dominate the oceans of the world.'

Cavitation bubbles are one of the reasons it's hard to make ships go fast: churning propellers create spirals of cavitation bubbles that wear away their surfaces. 'It's impossible to prevent cavitation above certain speeds,' Patek says. Cavitation also troubles submarine designers looking to build stealthy craft. 'If you're moving through the water fast enough, cavitation bubbles form, they collapse and it's deafening,' Patek says. 'It's a huge source of noise and a huge source of annoyance for people who work on undersea warfare and devices.'

Little smasher

The harlequin mantis shrimp is smashing at killing crabs. But this brings its own problems. When the shrimp hits the crab with a force of 700N followed by a 1,500N blow from a cavitation bubble, Newton's third law kicks in. Every action has an equal and opposite reaction. So as well as the shrimp hitting the crab, the crab is effectively hitting the shrimp back with the same amount of force. How does the shrimp avoid destroying its clubs through the battering of the tens of thousands of mealtimes before it next moults? The animal replaces its exoskeleton every three to four months, which is just as well because by then it may have shattered the covering on its dactyl 'elbows'.

Time to hear more from David Kisailus, the researcher who introduced us to the harlequin mantis shrimp. He's discovered that, as well as the two types of physics we heard about just now, the shrimp employs materials science. Its dactyls have a unique structure that delays the point at which they crack up completely by letting them crack a little. According to Kisailus, humans try to make things so strong that they don't fail, but biology does something a little different. 'Biology builds things that are both strong and tough,' he says. 'The dactyl clubs of the mantis are able to withstand multiple impacts by forming nano-scale cracks. Those cracks have to travel an incredible distance, constantly being redirected and twisted at less stiff interfaces.' During this journey, the cracks lose their energy. 'If you have a ceramic plate and you drop it on the floor, it will fracture completely,' Kisailus adds. 'But if you were to surround that ceramic plate with foam rubber, it would not crack because the energy would be absorbed by that softer component.'

To crack the secret of the dactyl's own shell, let's start at the outside. The striking face of the club has a hard outer edge made from particles of hydroxyapatite, a type of calcium phosphate similar to a mineral in human bone. The layer is just 60 microns (0.06mm) thick, equivalent to the width of a thin human hair. It must be hard so that it can smash the shell of the crab or snail under attack. On its own it would crack when beaten repeatedly onto shells, so behind it there's material that stops cracks in their tracks, as well as providing support and stiffness.

It's this material that is unusual – it has a Bouligand structure, named after French mathematician Georges Louis Bouligand (1889–1979). It consists of more layers, sheets of a complex sugar called alpha-chitin. The sugar molecules are long, like fibres, and line up next to each other. Each sheet is parallel to the surface but its lines of sugar molecules are rotated compared to the layer above, creating spirals down through the layers. In the harlequin mantis shrimp the fibres twist through 180°, that is from noon on a clock face to 6 p.m., over a depth of 75 microns (0.075mm). Surrounding the soft fibres lies a harder material: hydroxyapatite, as found in the striking face.

This layer beneath the hard outer edge is a composite, like rum and raisin chocolate, with softer pieces of fruit (the fibres) embedded in a harder matrix of chocolate (the mineral). This mix of compliant and stiff gives the material its toughness by deflecting any nano-cracks formed on impact around the spirals of the Bouligand structure. 'There's a large surface area of interfaces [between soft and stiff regions], which cause a crack to have to go through this long tortuous path,' says Kisailus. 'So the crack runs out of energy long before it can get through the club.'

Layers of this Bouligand structure continue down through the impact region. Further into the club, the make-up changes – the alpha-chitin fibres in the Bouligand structure sit in a mix containing a non-crystalline form of calcium phosphate, not in crystalline hydroxyapatite. What's more, the periodicity (thickness) of these layers changes, moving from thick to thin as you head further from the dactyl's hard outer edge. This gradation, Kisailus says, may play another important role: filtering acoustic waves during impact and protecting the material from shock.

Shelling out

The harlequin mantis shrimp is not unique in having a shell with a Bouligand structure; other arthropods (a group of invertebrates that includes insects, spiders and crustaceans) have this arrangement too. But this crustacean is special because its shell contains regions of both crystalline calcium phosphate and non-crystalline calcium phosphate and carbonate, and because its stiff (mineral) and soft (fibre) composite structure gradually changes in periodicity.

The shrimp could even teach us a thing or two. Kisailus has copied the design of the dactyl's energy-absorbing region. 'Instead of using biological materials like complex sugars and calcium phosphate, we used engineering materials like carbon fibre and epoxies and basically mimicked the architecture,' he says. The team made panels that showed almost half as much damage under impact as those used in the Boeing 787 plane. Not surprisingly, Kisailus is in demand from aircraft companies.

While Kisailus's studies take place in a laboratory, he dreams of fetching the species himself. The crustaceans are 'in some really cool locations like Vietnam or the South Pacific'. Instead, collectors send him shrimp that he keeps in separate tanks within an aquarium holding almost 2,000 litres (440 gallons). The tanks are made of tough plastic, as the harlequin mantis shrimp can smash its way through glass with its powerful club. It's pretty much the Houdini of the animal world. Kisailus has already noticed small cracks in the plastic tank of one particularly feisty mantis shrimp and says he may have to find tougher tanks still. Fortunately for the researchers, the shrimp's powers even extend to regenerating body parts. 'We can pull off one of their clubs and they'll regrow a new one, which is good,' says Kisailus. 'You kind of get attached to them after a while, they're cool animals.'

Fastest loser

Officially cool, the harlequin mantis shrimp is top notch at not smashing itself to pieces. It uses physics to attack its prey – Newton's second law along with power amplification to provide the acceleration for a strong blow, and cavitation bubbles for an extra punch – and materials science to protect its own body from cracking when Newton's third law kicks in. How does the shrimp shape up against other animals when it comes to moving fast? As we heard, the harlequin mantis shrimp has a linear acceleration as high as $100,000 \text{m/s}^2$. Although not the best performer in the animal kingdom, it's far up the table; if you were playing animal top trumps the harlequin mantis shrimp card could be a winner, especially since the creature's eyes also detect circularly polarised light (turn to Chapter 6 for details of polarisation). As far as we know, only one other species of mantis shrimp can do this. Although to guarantee a top trumps victory you'd need to leave certain animals out of the game.

'The highest documented acceleration in an animal so far is $1,000,000 \text{m/s}^2$, for the stinging cells of jellyfish – nematocysts,' says Patek. 'It's not a muscle-based system but it has exactly the same principle. It's a stinging cell wound with elastic fibres that get stretched like a balloon. And there's a little trigger hair that releases it. Again it's a very slow winding of energy and a fast release.'

In the animal kingdom, there's fast and then there's fast. It depends how you define it. Even then, there are arguments. A fast movement could take place very quickly, over fractions of a second; or have a high speed; or have a large acceleration. According to Patek, duration is the best way to define fast. Here jellyfish nematocysts win again: they use high-pressure capsules to fling out their poisoned stingers in around 0.7 milliseconds. Harlequin mantis shrimp take a tardy few milliseconds to slam out their claws. But the sting in the tail for the jellyfish is that, because of this short duration, the nematocysts don't reach particularly high speeds despite their high acceleration – there just isn't time to build up a good pace. You win some, you lose some, stingers.

If we define our league table more carefully, cheetahs probably take the number one spot for peak sustained speeds, with diving falcons winning gold for peak unpowered speeds – they tuck in their wings and use gravity to swoop.

Nematocysts and fungal spores win for shortest duration. But we're ignoring fungi in this book – they're not animals and don't even have muscles; instead, surface tension pops the spores off their stems. More on surface tension, and how it can perform miracles, in the next chapter.

Shut your trap

Even if you define its competition class precisely, the poor old harlequin mantis shrimp isn't a winner on the acceleration front. Pipping the harlequin mantis shrimp at the post in the all-new 'acceleration of an appendage relative to the body through power amplification' category, are trap-jaw ants and termites, which snap their jaws shut with an acceleration of almost $1,000,000\text{m/s}^2$. As their name suggests, trap-jaw ants have massive jaws; these carnivores use their mandibles to stun prey. Up to a centimetre long (almost half an inch), the ants hang out in tropical climates like those of Costa Rica, South America, Australia and the south-east US. Recently *Odontomachus haematodus*, a species native to South America, started invading the US Gulf Coast and south-eastern states; who knows where these ants will end up as climate changes?

Trap-jaw ants feed on creatures that, instead of running away, use tactics like spraying a noxious chemical. In self-defence, the ant stuns its victims with a quick blow from its mandibles. In the lab, Patek, who researches ants as well as shrimp and, as we'll find out in Chapter 4, lobster, feeds baby crickets to *Odontomachus bauri,* another of the 70 species of trap-jaw ant. But trap-jaw ants are feisty and they'll eat whatever you provide. 'They'll give it a shot even if it's huge,' Patek says. 'We sometimes give them mealworms and they just go up and hit it until it stops moving. If you put your finger in the tank, they'll come up and hit you. Nothing seems to put them off.' The ants look like they mean business – they have a bulbous head, chest and abdomen and long, spider-style legs, as if they've been made from Meccano (or Erector if you're in the US). They hold their mandibles, which are a couple of millimetres long and have a pincer at the far end, out wide, like the two halves of a handlebar moustache.

As they snap shut, the jaws of *Odontomachus bauri* reach up to 230km per hour (143mph) within 0.13 milliseconds. Like the mantis shrimp, these speedy snappers use power amplification. 'You can't do it otherwise,' as Patek puts it. But unlike harlequin mantis shrimp, which hold their dactyls closed then spring them open, the ants latch their jaws open against the power of their massive mandible closer muscles before snapping them shut. They bend their entire head to store elastic energy. 'When they're ready to strike they have a little muscle that moves the latch out of the way and allows the jaws to close,' says Patek. 'One of the cool features is that they have sensory hairs on the inside of the trap jaws; these hairs have neurones that go straight to the trigger muscle. It's an exceptionally fast trigger response as it doesn't have to be processed by the brain.'

This fast-accelerating snap produces a force that's hundreds of times the ant's body weight. It's so large that controlling it can be a problem. Some species of trap-jaw ant appear to use their jaw to move, banging it shut against the ground to send themselves cartwheeling into the air, perhaps to escape danger, to mob a predator to scare it away, or to spring back from a prey animal that's defending itself. Or it could just be a rebound effect – it's not clear how much control the insect has over these antics. Sometimes, Patek says, it seems to be 'sort of a blooper where they strike and then bounce backwards'.

Even if it's doing this by mistake, the ant's bouncebacks provide us with a good opportunity to introduce Newton's first law of motion (that's the last one for us to deal with; Newton's laws, unlike the laws of thermodynamics, don't have any prequels). Also known as the law of inertia, Newton's first law says that if the net external force on an object is zero then the velocity of that object will be constant. Without an overall force, something that isn't moving will stay that way. So the ant continues to sit on the ground until it springs its jaw shut against the Earth, whacking the ground with a strong force. The Earth acts in accordance with Newton's third law and pushes back on the ant with an equal and opposite reaction. And this force acts in line with Newton's first law to send the ant flying – it's a net external force that changes the ant's velocity.

Equally, under Newton's first law, an object that is moving will keep going with the same speed and in the same direction until a force interferes. Our Moon would head off into space in a straight line if the Earth's gravitational field didn't exert a pull to keep it orbiting around us. Newton's first law also gives us the modern definition of force, as any external effort that causes an object to undergo a change in its movement, direction or shape. Even the ancient Greek philosophers Aristotle and Archimedes had the hang of forces to a large extent. They knew that if you apply a force to something that's not restrained, it will move. But they were hampered by their lack of under-standing of friction – the force arising as two surfaces or layers of fluid rub against each other (we met friction in Chapter 1, where it heated up bored cannon). So these eminent thinkers believed that to keep something moving, you have to carry on applying a force, like a horse pulling a cart. The Greeks didn't realise that other forces act against the motion of the cartwheels: the friction of their outer rims with the ground, and the friction of their inner rims as they turn on the cart's axles. Under ideal conditions, on the other hand, such as an imaginary perfectly smooth, friction-free stretch of ice, an ice-hockey puck will slide at constant speed in the same direction without the need for any extra force until it hits the wall at the edge of the rink. Almost ideal conditions can be seen in space, for example in the film *Gravity* when actor Sandra Bullock floats gratuitously around in circles in her vest and pants, her rotational velocity constant in the absence of any net external force, and the film crew apparently, for the sake of aesthetics, wilfully ignoring the fact that astronauts wear nappies.

Back to reality. Newton's first, second and third laws team up to provide the trap-jaw ant with a strong gnash and movements it can't always regulate. There's nothing, though, to suggest that the gecko isn't in control as it strolls upside down across ceilings. Even if it falls it can grab a surface with just one foot and support its own weight. Small forces, acting together, can be strong too.

A sticky customer

As you amble down the hotel corridor back to your room after dinner, the Mauritius heat still warm on your skin, lizards with round eyes and mottled bodies scuttle across the ceiling, one per lamp fitting. They're much smaller than Komodo dragons and you're not in any danger. These geckos are hunting insects drawn to the artificial light. But how do they defy gravity and run Spiderman-like across the ceiling?

And how do they stick on tree trunks and beneath leaves in the rain? For geckos live in the tropics, where it's often wet. It was just this question that Alyssa Stark of the University of Louisville, US, got stuck into for her PhD while she was studying at the University of Akron, US. It made a change, as she began her career studying how eating the toxic algae produced during a 'red tide' harms memory in sea lions. An absent-minded sea lion may forget it has already met another individual and act aggressively until the two re-establish the hierarchy by agreeing which of them is the alpha animal. 'When I first started, the biggest difference was I could grab a gecko with my hand,' Stark says. 'I was used to sitting on top of animals, using all my body weight.'

The tokay gecko (*Gekko gecko*) is no stranger to aggression itself – it's one of the most feisty species. 'That's easier to

work with than those that are skittish and run away,' says
Stark. 'These guys just sit there and kind of bark at you and
try to bite you but it's easier to deal with than the running.'
A fleeing gecko has a lot more options than a human – it
can scale the walls and scamper across the ceiling. The
research team has a pole with a loop of cord on the end for
gently retrieving the ones that get away, but there are parts
of the lab that even the pole can't reach.

Normally geckos don't live in labs; Stark headed out to
Bali to look at the tokay in the wild. 'They make a call that
sounds like "tokay" – that's where they get their name
from,' she says. 'I could hear them calling and I could
pinpoint some of them but there were some that I couldn't.
They are usually close to a high spot and right when you
come by, they hide.'

Sticking around

Before we deal with how geckos cope with wet surfaces,
let's find out how they hang on to dry ones. Yup, it's all
about physics. As experiments by Kellar Autumn of Lewis
and Clark College, US, showed in 2002, the gecko uses van
der Waals forces, the tiny attractions between molecules.
Named after Dutch physicist Johannes Diderik van der
Waals (1837–1923), these forces are like a mini-gravity.
The science gets scary but basically the forces arise because
each molecule contains electrons whizzing round in random
orbits. These charges set up electric fields that can temporarily
attract another molecule close by.

Since they're due to weakly charged electrons, these
forces are weak too. Typically, they might provide an
adhesion energy – a measure of how much the molecules
want to stay together – of 50–60 millijoules (mJ) per square
metre. So how do van der Waals forces counteract the
body weight of a 100g (3.5oz) gecko? There's another snag:
van der Waals forces only work over distances of less than
10 nanometres. That's smaller than the size of a virus, so

the gecko must plonk its feet right up close to the surface it's trying to cling to; the molecules in its skin and the ceiling must be near enough to attract. To achieve this, species like the tokay gecko have fleshy folds known as 'lamellae' covering the soles of their feet. The result is like a rubber tyre tread, with folds spanning the width of each of the toes. In spite of its fleshy feet, the tokay gecko is beautiful, with large eyes with a vertical slit pupil, and a pale blue or grey body speckled with a mix of yellow, orange or bright red spots, as if decorated by a pointillist on acid.

Coating each lamella are tiny hairs known as 'setae' that are roughly 5 microns (0.005mm) wide and just over 100 microns long. 'You have a single hair with multiple branch points,' says Stark. 'They terminate in flattened spatulas; these tips not only allow the geckos to make really close contact with the surface they're sticking to but also a lot of contact.' Each spatula is typically triangle-shaped, about 0.2 microns wide and 0.2 microns long; each seta (hair) may contain between a hundred and a thousand of these tips.

A tokay gecko has a grand total of around 6.5 million setae, with about half a million on each foot. In theory the 2–20 square microns of spatula area for each of these hairs provides an adhesion force of about 40 millionths of a newton (40μN) if you pull the seta away from the surface, and around 200μN if you slide it. Every little helps. There are so many hairs, each with so many spatulas, that the total area of close contact between the gecko and the surface is big. And the small van der Waals forces that act where the molecules in the gecko hairs and the surface are right next to each other add up too. In tests, geckos could support a sliding force of 20–40N. So the gecko can support its body weight, which is 1N for a 100g (3.5oz) gecko, with just one toe. But this 20–40N is below the theoretical value of 1,300N. 'This is probably because perfect contact of all setae at the same time is unlikely,' says Stark. 'At the whole-animal level they are only using 3 per cent of their setae ... which is fine because to support their body mass they need less than 0.04 per cent!'

It's all done without glue. 'There's nothing about chemistry that is happening,' says Stark. 'It's just two surfaces coming into really close contact and forces being stronger at that close contact.' The gecko expert compares it to interlacing the pages of two telephone books, page on page, then trying to separate them. 'You have not glued them together, you haven't stapled them, you haven't done anything in particular to make them adhesive, but if you do that you can't pull them apart because you've got such close contact and so much of it,' she explains.

The spatula-tipped branching-haired feet also mean it doesn't matter whether the surface the tokay gecko is sticking to is rough or smooth. The hairs bend around lumps and bumps or squeeze into holes and gaps. Since van der Waals forces work between any molecules, it doesn't even matter what the surface is made of. As Stark puts it, 'generally you can run a gecko on pretty much anything.' There's one notable exception. The tokay gecko can hang on to almost any material except dry Teflon. It's non-stick, even for geckos.

What a feeling

Walking on the ceiling is a great party trick but insects are available at ground level too. Why did the tokay and some other gecko species develop such complex feet? The answer is simple: being able to run up vertical surfaces like tree trunks and walls and hang upside down from branches, leaves and ceilings opens up habitat for these geckos that other lizards can't reach, helping them to access new prey and avoid predators. 'They can get up and they can be safe when they do it,' says Stark. 'They don't really fall all that much and people say even when they do fall they can grab one hand on something and catch themselves. They can support a lot more than their body weight, there's this huge safety factor.' Over time, some gecko species developed, then abandoned, a sticky foot structure several times, perhaps as their living conditions varied and its usefulness to them waxed and waned.

If the adhesion is that strong, once a tokay gecko has attached its foot to a surface, how does it remove it? 'You don't want to be so sticky that you're stuck and you can be eaten by a predator,' says Stark. 'So it is important for them to be able to release it too.' The gecko has extra-bendy toes that it peels backwards, away from the surface, starting at the far end. For us, the equivalent would be placing our hand palm upwards on a surface and curling our fingers inwards to make a fist. This 'unzips' the spatulas and their adhesive forces one by one, which is easier as each force by itself is small. 'Geckos outperform the couple of other types of lizards that stick probably because of their ability to roll their toes backwards,' says Stark. 'They peel all the way off so that their toes are pointing up or even back towards themselves.'

Gecko toes get a little unpeeling help from their hairs. 'Most of those hairs are oriented at about a 30° angle,' says Stark, 'so that the gecko is sticky in only one direction.' Once the hairs hit the critical peel angle where the van der Waals forces are no more, they unpeel very easily and pop off, freeing the toe pad. 'All they have to do is barely lift their toes off and that contact is gone,' says Stark. 'So now you have an adhesive that you can reuse repeatedly really quickly.'

Singing in the rain

Time to get back to Stark's PhD mission: what happens in the wet? The way gecko feet work presents the reptiles with a sticky – or rather, potentially non-sticky – problem when surfaces are covered in liquid. 'You've got these forces that require close contact and if even a thin layer of water is between the toe and the surface, van der Waals won't work at all,' says Stark.

In the lab, scientists have generally looked at geckos under dry conditions. Stark focused on gecko adhesion when everything is covered in water droplets. Her initial tests revealed that geckos struggle to stick on wet glass. This seemed strange, given that geckos encounter water so often in their everyday lives. 'We took a step back and thought about what we tested them on,' she says. Instead of putting the geckos on hydrophilic, or water-loving, glass 'which is what most everybody uses because it's easy to get', Stark tried a water-hating – hydrophobic – surface. 'They stuck just as well [on that] in water as they did in air,' she says. 'The surface chemistry in a wet environment became really important whereas in air it's not important at all – van der Waals forces are basically surface insensitive.'

So how does a gecko stick to a wet hydrophobic surface? The answer lies in what happens when you try to soak one of the reptiles. 'When you squirt a gecko with water, the water just flies off,' says Stark. 'It's really hard to wet a gecko. The toe-pads are superhydrophobic and you can see the water beaded on top.' In extremely wet conditions, the same property that makes the water bead up creates small pockets of air around gecko feet, according to Stark. 'If you push a foot that's really water-repellent through water, you've now got an air bubble around that material so that when they press down they stay dry and can make that close contact still.'

Stark found that tokay geckos can run well on a hydrophobic plastic surface coated with water droplets as if soaked by a rain shower. The animals travelled 2 metres vertically on a wet surface just as fast as on a dry one. Even if a gecko gets its feet so wet that they no longer stick, it can dry them by repeatedly peeling and unpeeling them onto and off a surface, in the same way that it would 'self-clean' away dirt. 'If they're allowed to peel they can dry that toe-pad faster than if they were just sitting and letting it dry by evaporation,' says Stark.

On hydrophilic (water-loving) glass, however, this air-bubble system doesn't work. And oddly, Stark discovered that tokay geckos can stick to Teflon only when it's wet. 'I didn't even want to do that experiment,' she says. 'I had an undergraduate student who persuaded me to do it for fun and then I was shocked when it not only worked but was better. We're still not sure if we can explain this.'

Hang on

It's no surprise that these super-sticky feet have caught the eye of scientists developing strong, reusable adhesives. But the gecko is better at making gecko feet than we are. 'My advisor before I started my PhD, Ali Dhinojwala at the University of Akron, had designed a synthetic gecko tape made of carbon nanotubes,' says Stark. 'It was stickier than the gecko, it could self-clean, it was hydrophobic, it was reusable, but it had some issues – if you sheared it in a particular way the carbon nanotubes would kink and not come back. The little hairs on a gecko toe are resilient and they pop back when you smash them.'

Even though we can make hair-like structures from many polymers, we can't yet capture all the properties of the gecko. 'It's making it difficult for us to move forward towards where you can have a Spiderman hand that can support your body weight,' Stark says. 'We're doing a lot of investigation on the animals right now, because we're a bit stuck on the materials side.'

That said, in 2014 the US Defense Advanced Research Projects Agency (DARPA) announced that under its Z-Man project, a 100kg (220lb) man had climbed an 8m-high (26ft) glass wall with the aid of two hand-held paddles coated in 'Geckskin'. Described in a scientific paper in 2012, this material is based on a stiff fabric coated with a soft layer of plastic, rather than hairs, that increases the amount of contact area with a surface. It was developed by the University of Massachusetts, Amherst, and made at the Draper Laboratory in Massachusetts, US. 'What is neat about "Geckskin" is that it doesn't take the obvious approach to gecko-inspired synthetic adhesive design,' says Stark. 'Most designs make hair-like structures on a backing material. Here they focused on the backing material and used a smooth adhesive contact. Their inspiration was not the hairs on the gecko toe but the tendons which connect the toe-pad lamellae. Out-of-the-box thinking like that can move the field in new directions.'

Sticking point

Gecko feet are keeping secrets from us. But Stark has discovered a clue. One day, she and her colleagues noticed something else strange. Something that made the problem stickier still. 'The lab happened to be dark,' she says. 'We turned the light on to clean a piece of glass that we had a gecko walk across and we noticed a little bit of residue in the shape of a footprint. We looked into it further and we realised it was a lizard footprint.'

The residue was made of lipids, natural oils also found on gecko skin and in many other types of animal. This was weird as, until Stark's discovery, researchers had thought gecko feet stuck solely through van der Waals forces, without any glue-type action. 'The gecko is touted as being a dry reversible adhesive that doesn't leave any residue,' says Stark. 'Technically that's not true any more. Even though you can touch a gecko foot and not notice any lipids being left on you, they probably are.'

It's not yet clear whether lipids play a role in giving geckos sticky feet, but there are a number of theories. The lipids could help adhesion, for example, or keep the gecko's foot hairs clean, or make them more waterproof. 'They're leaving something behind when they're sticking so maybe without that it would be better or worse,' says Stark.

'We don't really know.' Understanding what's going on with gecko foot lipids could help scientists make their own sticky materials. 'The gecko adhesive system has a lot of great properties that we can't replicate yet,' says Stark. 'These lipids might be a clue to any one of those properties. We really think there could be something interesting there.' We leave you with residue.

Snap-shot

So the Komodo dragon compensates for its feeble bite force with a strong neck and body, and mosquitoes play with momentum and impulse to battle damage from raindrops. The harlequin mantis shrimp uses high acceleration to batter crabs with a large force according to Newton's beautiful second law (F=ma), and receives pushback from their shells in line with Newton's third law, protecting itself with a clever 'bone' structure. The trap-jaw ant uses the same acceleration technique to stun its chemical-spraying prey, again with power amplification releasing the blow faster than the animal's muscles could without it. Geckos, meanwhile, add a lot of small van der Waals forces together to stick themselves to the ceiling. They can even hang on when it's wet, leading us seamlessly to the topic of our next chapter: fluids.

CHAPTER THREE

Fluids: When Things Get Stickier

POND SKATERS THAT WALK ON WATER * CATS DEFYING
GRAVITY * STEALTH SEAHORSES * CONVENTION-DEFYING
BEES * PTEROSAURS ON THE EDGE

Walking miracles

If you couldn't swim, would you try learning from a book?

Most of us wouldn't take the risk. We'd head for the swimming baths for lessons with a qualified instructor and a plastic float. But German theoretical physicist Theodor Kaluza (1885–1954) wasn't one to bother with anything boring like that. Best remembered for his madcap ideas about a five-dimensional world, Kaluza was so convinced by the power of theoretical knowledge that he reckoned a book on swimming would teach him all he needed. So one day in his thirties, after boning up on the background, Kaluza jumped into some water. And swam, at the first attempt.

Details of his feat are sketchy, mainly relying on an account by Kaluza's son, but that didn't stop the scriptwriters of *The Big Bang Theory* shoehorning a version into the hit TV show. Watch series 2 episode 13 and you'll see geeky physicist Sheldon (played by Jim Parsons) explain how he learned to swim by reading up on the Internet. Flatmate Leonard (Johnny Galecki) is unimpressed when Sheldon admits he only managed to swim on the floor, but Sheldon insists it was useful. 'The skills are transferable,' he explains. 'I just have no interest in going in the water.'

Sheldon is right to be cautious: water can be dangerous. If Kaluza hadn't been able to swim, the downward pull of gravity on his body would have outweighed the upward buoyancy force of the water, which equals the weight of water displaced – as Greek scientist Archimedes realised in the third century BC. And the theorist would have sunk, leaving his five-dimensional thoughts lost to posterity. Since he could swim, Kaluza used his muscles to push water down and backwards, creating a force in the opposite direction as per Newton's third law (see Chapter 2). This upwards and forwards force stopped him from sinking and moved him along.

Some animals, however, live around water yet don't need to swim. Crouch near the surface of a lake and you might – if you stare hard enough – see an insect walk on water. This animal is the pond skater, of which there are more than 1,700 species. With a thin, brown or black body about 1cm (0.4in) long, the pond skater has six slender stick–like legs splayed out, three on either side. It can scoot across the surface of water at well over a metre per second (about two miles per hour).

Examine a pond skater sitting still and you'll see the lower part of its legs lying flat on the surface of the water, making tiny dents like a bowling ball on a mattress. This ability to sit on the surface, and not sink, is handy for a pond skater eyeing, say, a spider that's landed nearby. Too heavy and with the wrong kind of legs, the spider can't walk on water so the pond skater scurries over, grabs the arachnid and punctures it with the claws on its front feet, before sucking out the goodies inside.

Despite this ghoulish behaviour, pond skater names are matter of fact. They're known as water striders, water skimmers, water skaters, water scooters and water skippers. They're an elite bunch; just 0.1 per cent of insects can walk on water. Some people even call them 'Jesus bugs', though we don't need miracles to explain why they don't sink. Just simple physics and a spot of knowhow about the theme of this chapter: fluids.

Fluid thinking

When scientists talk about fluids, they mean both liquids and gases. Sometimes it can be hard to tell the two apart. Both flow. Both adopt the shape of their container. Neither bounces much if you push it. You could say the difference between a liquid and a gas is rather fluid. There are a few differences though. You can compress a gas (think pumping up a car tyre), but not a liquid. Liquids are also much denser than gases: a litre of orange juice weighs about a kilogram, but a litre of air weighs barely a gram. And a liquid has a surface, whereas a gas doesn't.

We'll return to gases later to see how bees and pterosaurs fly, but – like our pond skater – first we need to get to grips with liquids. Water, to be precise. It's the most abundant liquid on Earth, covering over 70 per cent of the planet's surface. And it's a remarkable substance despite its simple structure: a collection of V-shaped molecules each containing just one atom of oxygen connected to two of hydrogen, giving the chemical formula H_2O. Water's almost an 'ideal' fluid as far as physicists are concerned – not because they like drinking it better than anything else, but because the way water flows is about as simple as you can get. No matter how slow or fast you stir water, its viscosity is the same. Our old friend Isaac Newton gets the credit for the first serious study of such fluids, which are known in his honour as 'Newtonian'.

Not all fluids are Newtonian. Tomato ketchup becomes runnier the harder you push it. That's why it oozes freely from a plastic container, which you can squeeze, but is so damned hard to dislodge from a rigid glass bottle (forcing you to smack it out with your palm until your hand hurts). Another non-Newtonian fluid is hair gel. It's 99 per cent water, yet you can't pour it, though it's best not to debate the point with airport security staff if you want to avoid missing your flight. For the record, you're banned from sneaking on gels – hair, hand, lip or anything else – in your carry-on luggage on a plane. Ditto toothpaste and shaving foam, which are also non-Newtonian fluids.

Animals don't have to worry about ketchup or hair gel, so let's get back to water. We said it was an ideal fluid, but that's only regarding how it flows. In other respects, water's weird. Just think about its density. Solids usually become less dense when they melt as their atoms or molecules break free from the rigid lattice keeping them in place, and start to drift apart. Liquid water, however, is *more* dense than ice, which is why icebergs float, as passengers on the *Titanic* discovered to their cost. Water also conducts heat faster than any other common liquid, which is why you cool so quickly when you jump into a cold swimming pool (see Chapter 1). And water has a massive 'heat capacity', meaning it can absorb lots of heat without warming much; think how long it takes to boil a pan of the stuff on a hob. Especially if you watch it.

But none of these properties aids our pond skater. The clue to the pond skater's skill lies in water's bouncy, trampoline-like surface. Deep inside a liquid, any molecule is surrounded on all sides by other molecules, which attract it equally in all directions. But a molecule at the surface has fellow molecules only below; above the surface there's just air. It therefore experiences a net downward pull, and overall the whole surface becomes taut as if covered by a stretchy elastic sheet. This stretchiness, or 'surface tension', is found in all liquids, but water has a higher surface tension than almost any other. At 73 millinewtons (mN) per metre at 20°C, it's beaten only by liquid mercury – and that doesn't occur in nature. Somehow the pond skater exploits water's bouncy surface not only to avoid sinking but also to scoot across this most special of fluids.

A tense affair

Back in the early 2000s, John Bush, a fluid dynamicist at the Massachusetts Institute of Technology (MIT) in the US, along with David Hu, who later carried out the shaking-dog experiment in Chapter 1, and Brian Chan calculated the forces on a pond skater sitting on water. Basic physics told them that the insect can sit on the surface of a liquid only if its weight (the downward pull of gravity) is less than the force from the liquid pushing it upwards. If it weighs more than the upward force, it'll sink. Which means that for a pond skater to live up to its name, it needs to make that upward force as big as possible.

So what are its options? As Archimedes pointed out, the upward force on an object in a liquid equals the weight of water displaced. But for something on the surface of a liquid, different rules apply. There the upward force depends on three things: the liquid's surface tension, the object's length and the maximum angle it dents the surface of the water below the horizontal. To win most support, a pond skater needs all three quantities to be as big as possible. The animal can't adjust the surface tension of water, which is a fixed value, though it helps that it's already so high. Instead, the secret to the pond skater's water-walking ability lies in its six long, spindly legs, which are hinged at the middle. The ends sit horizontally on the water, like a set of personal water-skis. On most pond skaters, this portion of the limb is 1cm (0.4in) long, boosting the total upwards force on the animal to a value well above its total weight. Result: no sinking. It's good to have a safety margin, though. The largest pond skaters, *Gigantometra gigas*, are near the limit. With a mass of roughly 3g (0.1oz) – about a thousand times heavier than the smallest species – they have legs measuring more than 20cm (8in) to generate enough upwards force to stay above water. When it comes to long legs, they're the Daddy.

Get your skates on

So surface tension means pond skaters keep their heads – and the rest of their bodies – above water. But how do they skate across the surface? To move forwards you have to push backwards on something – it's Newton's third law again, which says that if you put a force on a body, it'll exert an equal and opposite force on you. Humans and other big, land-based animals find it easy to propel themselves forward as the ground's solid. Swimming or rowing is tougher because your legs or oars have to move a pile of water backwards to gain forward momentum. Plus the water's much gooier than air so it's harder to move through. A pond skater, however, can only push back against the surface. So how does it get the oomph to move? And how does it move forwards rather than up?

To find out, just look at a pond skater skating and you'll see it leave tiny ripples on the surface of the water in its wake. Scientists used to think these 'capillary waves', formed by the animal's legs, are what propels the insect forward. But in 1993 Mark Denny, a biologist at Stanford University in California, noticed something odd. To make these waves, a pond skater must go quicker than about 25 centimetres per second – the lowest speed that a wave travels on a liquid surface. Big, long-legged pond skaters can certainly go that fast, but infants move much more slowly. So how come these littluns walk over water too?

'I read about Denny's paradox in a book by him, and thought I could sort it out,' Bush recalls, referring to the puzzle of the mysteriously moving infants. So in 2003 Bush, Hu and Chan plucked some pond skaters from local ponds. The animals reproduced every few weeks, giving our trio the perfect opportunity to film baby pond skaters scuttling over the surface of the water. Placing the youngsters in a small aquarium, they used high-speed video cameras to film the animals' antics, adding food dye in the water to see how the liquid moved.

To the researchers' surprise, the dye revealed that day-old pond skaters move by using the middle of their three pairs of legs like oars on a rowing boat, just like the adults. But they're moving too slowly to create capillary waves. Instead, as they scull, the insects shed vortices – small swirls of water that move backwards beneath the surface. A vortex is a region of low pressure in a fluid; you can make one by letting your bathwater race down the plughole or using a knife to stir a cup of creamy coffee so that the liquid rolls up and around the edge of the blade. But unlike fish, which swim by sending away spherical vortices, pond skaters make hemispherical vortices. The flat part of each hemisphere, which is about 8mm (0.3in) across, lies parallel to – and just below – the surface of the water, with the dome underneath. As these vortices move backwards, their momentum is enough to propel the insect forwards, like a rocket surging ahead by ejecting hot gas. Pond skaters are so efficient that they can move 10–15 body lengths on a single stroke. That's like the rowers in a 20m (66ft) eight-person racing boat travelling up to 300m (nearly 1,000ft) every time they move their blades. Despite Denny's calculations, Bush's experiments showed that capillary waves do help to move pond skaters – both infants and adults alike – a little, but it's to a much smaller extent than the vortices. Again, it's like being in a rowing boat. You make waves, but it's mostly the forward momentum you get by pushing water backwards that sends you on your way.

Intrigued by their observations, Bush and his two colleagues decided to mimic the animals' motion by building a larger-than-life mechanical pond skater. Dubbed Robostrider, this beast isn't as menacing as it sounds. Its 9cm (3.5in) body was made from aluminium cut from a fizzy-drink can, while its legs were fashioned from stainless-steel wire. The researchers powered Robostrider by running an elastic thread from a sports sock down the length of its body and connecting the elastic to each leg via a pulley. Weighing just 0.3g (0.01oz), Robostrider acted like a real pond skater. With its weight supported entirely by surface tension, this artificial creature moved forward, just like its real counterpart, by rowing its legs to make hemispherical vortices. It travelled about half a body length per stroke – roughly 18cm (7in) per second (about one-fifth the speed of a real pond skater). 'It moved relatively clumsily, like a water strider wearing chain mail,' says Bush. Though that's not bad for a beast made from a sock and a fizzy drink can.

The great Italian scientist, engineer and artist Leonardo da Vinci (1452–1519) once wondered whether we could walk on water by donning elongated floats on our feet and using two poles for balance. For Leonardo, who wasn't aware of surface tension, the idea got no further than a few sketches, but if you run through the sums, it turns out we would need shoes about a kilometre (over half a mile) in diameter to take advantage of water's stretchy surface. That's just not practical. Even the intrepid Kaluza wouldn't have been able to walk on water, no matter how hard he studied.

The cat on the lap

Pond skaters should rejoice that water has such a high surface tension. It stops them sinking and lets them propel themselves across ponds and lakes. But pond skaters are a superficial bunch. Most animals have a much deeper relationship with water. Fish swim in it, absorbing dissolved oxygen. Hippos lead a double life, grazing on land at night but spending their days in rivers, coming up every few minutes to breathe. Crucially, mammals and birds won't survive unless they can get water into their bodies. Even cats need to drink, despite their aversion to getting wet.

But have you ever stopped to think how you drink? For us, it's easy. Fill a glass at the tap, grab a coffee or pour an orange juice from the fridge. Lift the vessel to your lips and pour the liquid into your mouth. Obviously without making any disgusting slurping noises; only other people do that. We've even got two back-up techniques. First, we've got a complete set of cheeks so we can make a partial vacuum in our mouths when we suck in, which is how to sup a cocktail with a straw. The pressure in your mouth is lower than outside, with the difference counteracting the force of gravity and drawing the drink up into your mouth. It's like having your own personal vacuum cleaner. The other, rather revolting method is to put a tube in your mouth, connect a funnel to the top and ask a friend to pour the liquid in. Lean your head back and the pressure of the column of liquid forces the fluid down your gullet – ideal for students wishing to consume lots of beer as fast as possible in drunken drinking games (or so we've been told).

Other animals have to make do with fresh water, not beer. As it lies mostly in puddles, ponds, lakes or streams, they've developed a variety of strategies for supping. Pigs, sheep and horses are like us – they have complete cheeks and can drink by sucking the water up. Frogs absorb water through their skins, while the desert-dwelling Merriam's kangaroo rat (*Dipodomys merriami*) extracts water entirely from the food it eats, even if fresh rainwater is about. Hummingbirds dip their tongues into nectar, with the sticky fluid flowing up grooves in the tongue like ink moving through blotting paper. As for the Namibian desert fogstand beetle (*Stenocara gracilipes*), it lives in one of the driest places on Earth and collects water from the fogs that drift in off the Atlantic Ocean every morning. The beetle sticks its bottom in the air so its body is at about 45° to the ground and waits for the tiny water droplets landing on its back to clump together and roll down into its mouth.

But what about cats? How they drink is a question many scientists have overlooked in pursuit of supposedly deeper quests, such as searching for the Higgs boson or designing a pen that can write in space. Part of the problem is that a cat moves its tongue up and down so fast that it's impossible to watch what's going on with the naked eye. Only high-speed photography is up to the job, with the first-ever attempt to film a cat drinking being the 1940 Hollywood documentary *Quicker'n a Wink*. Featuring US electrical engineer Harold Edgerton (1903–90) demonstrating the prowess of his 'stroboscopic' photography, the nine-minute film bagged an Oscar the following year in the 'best short subject one-reel' category. Edgerton's technique captures fast-moving events on camera by flashing a light on and off at a speed of your choosing. If it flashes at, say, 2,000 times a second you can record a film at 2,000 frames per second, as long as your camera can take pictures that fast, which Edgerton's could.

Quicker'n a Wink wowed audiences by showing the first-ever slow-motion footage of fast-action events, including a popping soap bubble and a golf ball fired through a telephone directory. The film also reveals how you can make a rotating electric fan appear stationary by adjusting the flash so it fires on and off at exactly the fan's speed. There's footage too of an egg being dropped on the fan blades as they go round – surprisingly, it bounces a few times before smashing to pieces. Fun though it was, *Quicker'n a Wink* had barely 20 seconds of cat-drinking action and wasn't a scientific study.

The world had to wait nearly 70 years for the first serious research into how cats lap. It all began one morning in 2008 when Roman Stocker – an environmental engineer then at MIT – was watching his grey cat Cutta Cutta over breakfast. Intrigued by how the moggy was drinking, Stocker decided to bring a modern, twenty-first-century high-speed camera back from his lab and record Cutta Cutta lapping at his water bowl. It proved to be an inspired moment in the history of furry logic.

Lapping it up

Cats have cheeks like we do, but they're not complete, which means moggies can't close their lips around an object to suck liquids up. 'Cats are carnivores and need to get their jaws wide open to catch and eat prey,' explains Stocker. Like many land-based vertebrates, cats instead use their tongues to move fluid into their mouths. When he looked back at his videos of Cutta Cutta, Stocker discovered the cat's technique. The common cat (*Felis catus*), he noticed, first sticks out its tongue and curls the top surface back sharply at the tip. Next the cat lowers its tongue towards the water. But rather than slashing into the fluid, the cat briefly rests the curled-up tip of its tongue on the surface. Attractive forces between the water molecules and the top surface of the moggy's tongue make the liquid stick to it. The cat then lifts its tongue, drawing the water up to create a vertical column of liquid, which lengthens and thins as the tongue moves higher. Once its tongue is back in its mouth, the cat shuts its jaws, capturing part of the liquid column inside. It has to do all this before the force of gravity breaks the column up and the liquid drops back into the bowl.

So it's a three-step process. Touch tip of tongue on liquid surface. Lift tongue up to create a column of fluid. Snap jaws shut to trap some of the column. You can make a column of liquid yourself by touching the base of a spoon on the surface of a bowl of water and raising the spoon up fast. You'll briefly see a vertical column of fluid, though it won't go any higher than a few millimetres before breaking and splashing back into the bowl under gravity's pull. A cat can do better thanks to the water-attracting surface of its tongue, but it still can't counteract gravity entirely. That's why it has to lap so fast. After filming 10 adult cats, Stocker noticed that a moggy sticks its tongue in and out of its mouth about 3.5 times a second while drinking, each time capturing roughly 0.14 millilitres of fluid (barely a thirtieth of a teaspoon). Eventually, after anything between 3 and 17 laps, the cat swallows the liquid collected in its mouth.

It's tough being a thirsty cat, but life was about to get a whole lot tougher for Stocker, who found himself in a cat-fight of his own.

You little copycat

To explore in more detail what Cutta Cutta was up to, Stocker roped in his colleagues Jeffrey Aristoff, Sunny Jung and Pedro Reis to copycat the animal's drinking mechanism. First the team held a glass disc coated with a water-loving material – just like the tip of a cat's tongue – above a bath of water. Next, they wired up the disc to a motor, which they programmed so that the disc could move down to skim the surface of the water, before zooming back up, exactly like a cat's tongue. Thanks to this kit, the team could repeat the entire up–down sequence at will, moving the disc at whatever speed or to whatever height the researchers fancied. A case of kit being better than cat – even if, as Stocker recalls, the tests 'were much easier said than done'. Filming the experiments with a high-speed camera, the team saw that each time the glass disc rose, it lifted a column of water just as a cat's tongue does. And after roughly a twentieth of a second, the top of the column broke away from the disc and fell back down.

Using the data this disc set-up gave him, Stocker derived a mathematical formula describing how fast a cat laps in terms of the length and width of its tongue. The formula says the frequency is proportional to the square root of the height of the tongue above the water and halves as the width of a cat's tongue doubles. If a cat laps more slowly than this rate, then each time it puts its tongue back in its mouth, the column of fluid will be long gone and the moggy won't get much to drink. If the cat laps faster, it can't pull much water up in the first place, which is also no good for quenching its thirst. So the cat sticks its tongue in and out at the Goldilocks 'just-right' lapping frequency of 3.5 times a second, which gives the animal a decent column of fluid and is also quick enough to catch the column in its mouth before it breaks.

Yet cats don't let their secrets out of the bag that easily. When Michael Nauenberg, a physicist at the University of California, Santa Cruz, learned of Stocker's formula, which was published in the journal *Science* in 2010, he plugged real-life numbers into the equation to check the findings. To his surprise, Nauenberg found that if the formula were correct, a cat with a tongue 1cm wide lifted to 3cm above the surface of the water should lap more than 30 times faster than it does. So, Nauenberg argued, the formula was wrong. He claimed the fault lay in Stocker's assumption that the column of fluid the cat creates stays in place through surface tension and doesn't move. 'Their arguments were based on an unphysical assumption that the liquid column was somehow confined,' says Nauenberg. 'This is manifestly not the case.' As Nauenberg pointed out, the fluid continually drains away under gravity, narrowing the column as it goes. Take that fact into account, as Nauenberg did, and you end up with a different formula, which says that the ideal frequency for cat lapping is proportional to one divided by the square root of the tongue's height above the water and doesn't depend on tongue width at all.

Nauenberg found that his formula successfully matched Stocker's original measurement that cats lap at a Goldilocks frequency of about 3.5 times per second. In 2011 he published his analysis in a paper in *Science*, to which Stocker and colleagues wrote a response in the same issue. That's when things got catty. While Nauenberg agreed with Stocker's finding that a cat's frequency of lapping is dictated by the balance between gravity (the weight of the liquid column) and inertia (how hard it is to pull the liquid up), Stocker argued it made no sense for his rival to plug numbers into Stocker's formula as it wasn't a precise mathematical equation. Instead, it was a 'scaling law' – a mathematical generalisation, which in this case didn't say what the lapping frequency should be for any particular cat. 'Nauenberg interpreted a scaling law as an exact equation, which one cannot do,' says Stocker.

That hadn't stopped Nauenberg calculating a scaling law of his own – the formula mentioned above. It didn't impress Stocker. 'His scaling law is in fact no better than ours as we demonstrated in our response,' argues Stocker. Nauenberg, however, thinks that Stocker and co's comments 'indicated that they did not understand my objections', but Stocker's co-author Sunny Jung, who's now at Virginia Tech, insists, no – his side were right.

Like we said, it got a bit catty, even if such debates are the lifeblood of science. We're not going to adjudicate who won, we'll just say that both sides agree how cats drink: touch tongue on surface, lift tongue up, snap jaws shut to trap the liquid column. It was the mathematical detail behind the perfect lapping speed for a particular size of tongue that was in dispute. Mind you, if you think cats are complicated, let's not get started on dogs. They drink by sticking their tongues right into the water and scooping it out. Dogs mostly ladle, not lap. It's messy and it lacks the beauty and elegance of feline drinking. With those attributes being crucial for many physicists, who turn their noses up at explanations for natural phenomena that get too complex, we hereby declare that cats are a physicist's best friend.

Going with the flow

Since what goes in must come out, before we move on from drinking, there's one last issue. How long does it take a mammal to urinate? It's hardly the ultimate question of life, the universe and everything, to which the answer, calculated by the Deep Thought computer in Douglas Adams's novel *The Hitchhiker's Guide to the Galaxy*, was 42. Turns out, though, that the number of seconds a mammal takes to empty its bladder is always half that value. The discovery had nothing to do with Deep Thought – credit goes to David Hu of dog-shaking and pond-skater fame, who is now at the Georgia Institute of Technology in the US. After watching 32 different animals from 16 species urinating in videos on YouTube and in the flesh at his local zoo, he found that all mammals weighing between three kilograms and eight tonnes take 21 seconds to empty their bladder (give or take 13 seconds, which might seem like a big range but not if you look at the wide variety of animals he studied).

An elephant – the largest animal Hu looked at – has an 18-litre bladder, yet its urine takes the same time to gush out as it does from a cat with a bladder some 3,600 times smaller (rats, mice and other small animals expel urine as individual drops, so different rules apply). The reason why the time's the same is to do with the length of the urethra, the pipe that links a mammal's bladder to the outside world. Bigger animals have longer urethras: in an elephant it's about 1m (3ft) long, compared to just 5–10cm (2–4in) in a cat. As the downwards gravitational pressure on liquid in a pipe is proportional to the length of the pipe, the liquid flows out at a faster rate for elephants (long urethras) than for cats (short urethras). The faster rate exactly compensates for the bigger bladder (the diameter of the animal's urethra is only significant for those smaller animals that urinate in drops).

For their efforts, in 2015 Hu and his students received an IgNobel prize, which honours achievements that 'make people laugh, and then make them think'. Getting rid of urine pronto is, however, no laughing matter for an animal in the wild, which is defenceless while weeing. Fortunately, all large mammals – including humans – urinate just five or six times a day, meaning we spend only 0.2 per cent of our lives on the job. It's not only efficient, but gives us plenty of spare time for thinking great thoughts. Which is handy as there's one question about fluids that – if you racked your brain for long enough – could make you a millionaire.

Numbers game

Right, about that money. There's no catch, we promise. You won't have to gamble your savings on the stock markets or email your bank details to the lawyer of a deceased oil baron in Nigeria. There is one thing, though. You'll need to be a maths whizz and solve one of the hardest problems in science. So let's get started.

Up to now, we might have given the impression that everything about fluids has been sewn up. Surface tension? Tick. Heat capacity? Tick. Viscosity? Tick. While we do understand fluids well, there's one thing that still has even the brightest physicists scratching their heads. If you can show you truly, madly, deeply understand this property, there's a million dollars up for grabs. That's because in 2000 the Clay Mathematics Institute, founded by Boston businessman Landon T. Clay and his wife Lavinia D. Clay, launched a set of seven 'Millennium Prize Problems', each connected with fundamental questions that remained stubbornly unsolved despite years of effort. To encourage people, Institute bosses rustled up an eye-popping $7 million (almost £5 million) prize fund, offering $1 million (£700,000) in cash for each problem solved. At the time of writing, six of the seven remained unsolved – so get your thinking caps on for a chance of the lolly.

All but one of the remaining challenges concern abstract number problems so we can ignore them; they're pure mathematics, not physics. We're focusing instead on an equation named after two nineteenth-century geniuses: French engineer Claude-Louis Navier (1785–1836) and Irish-born physicist George Gabriel Stokes (1819–1903). The equation also concerns seahorses – pointy-headed, upright-swimming curly-tailed ocean-dwellers we'll come to shortly. Although Navier and Stokes never worked together, both were interested in how simple, incompressible fluids like water flow. The mathematics that emerged from their studies, known today as the Navier–Stokes equation, describes how fast, and in which direction, a fluid is moving, not just on average but at every point. Solve the equation and you get something looking like a wind map in a TV weather forecast: information on the speed and direction that the fluid's moving at lots of different places.

The equation is easy to solve if a fluid's moving nice and smoothly, as it does if you turn the hot tap on slightly to top up your bath. The water pours out in a gentle stream shaped like a cylinder. Yank the tap to full blast, however, and the water cascades out in an irregular torrent, with vortices where it's circulating. The Navier–Stokes equation can't properly explain this turbulent flow. Yes, you could write a fat piece of software and use a supercomputer to crank out the details of the water's swirls. That's what weather forecasters do when modelling how gases circulate in the atmosphere to predict if it'll rain tomorrow. But using a powerful computer to spew out an answer to the Navier–Stokes formula for a turbulent fluid doesn't mean you've solved it from first principles. It's mathematical cheating and isn't going to win you that Millennium Prize. Sorry.

But don't feel bad; turbulence has left many of the world's top minds stumped. As the British mathematician Sir Horace Lamb (1849–1934) said at a meeting of the British Association for the Advancement of Science in 1932: 'I am an old man now, and when I die and go to heaven there are two matters on which I hope for enlightenment. One is quantum electrodynamics, and the other is the turbulent motion of fluids. And about the former I am rather optimistic.' Back then, quantum electrodynamics, which describes how light and matter interact, was a new theory deemed impossibly avant-garde and difficult. Eight decades later, we have quantum electrodynamics licked, but turbulence remains an enigma.

Slowly does it

Turbulence is not just annoying because we can't fathom it scientifically. It's no fun either on a rocky plane flight. An unexpected bout of turbulence can be violent enough to judder the aircraft and spill your pasta and cheese sauce into your lap. If you're not strapped in, you could end up hitting the ceiling. To reduce the effects of turbulence, our tip is to stay belted up at all times and don't sit at the back of the plane, where the movement's strongest.

Turbulence at dinner time is also best avoided for the dwarf seahorse (*Hippocampus zosterae*). Despite their name, seahorses aren't horses, but fish that look like horses. The dwarf seahorse, at about 2.5cm (1in) long, is not even the smallest of the 54 species. That title belongs to Satomi's pygmy seahorse (*Hippocampus satomiae*), which is barely 1.4cm (0.5in) long. Dwarf seahorses do excel at one thing, though: they hold the *Guinness World Records* title for the slowest-moving fish, travelling at less than 1.5 metres (5ft) per hour. If there were an underwater 100m sprint for dwarf seahorses, they'd reach the finishing line in nearly three days. But their slowly-slowly approach has science behind it.

Living in the Caribbean, in the Gulf of Mexico and off the south-east US coast, the dwarf seahorse is whitish-yellow with a long, pointy head that looks like the knight in a game of chess. With a thin body ending in a pretty curled tail, it dwells mostly in calm, sheltered waters, particularly where fields of green seagrass carpet the ocean floor. It loves to dine on copepods. Once a dwarf seahorse has swum up to one of these millimetre-sized transparent crustaceans, it rapidly rotates its head, pointing its snout upwards. The seahorse then sucks up its prey through its mouth. However, it can only do this swivel-and-suck routine when the copepod is right up close – less than a millimetre away. That's because to avoid being sucked up, the copepod has antennae covered with hairs that detect fluid flow. Within 2 milliseconds of noticing something suspicious in the water, a copepod zooms off from the danger zone at more than 500 body lengths per second. Scaled up, that would be like an athlete running a kilometre in a single second. So to eat, a dwarf seahorse must creep up on the crustacean without noticeably disturbing the water. Turbulence is a big no-no.

But how do seahorses minimise their turbulence? Could it be that they've sussed fluid physics better than we have? To find out, Brad Gemmell, a zoologist at the University of Texas at Austin, along with his Texas colleague Edward Buskey and Jian Sheng from the University of Minnesota, blitzed dwarf seahorses with laser beams. This isn't as cruel as it sounds: the team used red laser light, to which most sealife isn't sensitive. Rather than working in an ocean, the three researchers filled a 4cm by 4cm by 4cm glass aquarium with seawater. Then they added copepods collected from the Gulf of Mexico and microscopic 'diatoms', single-celled organisms for the copepods to feast on as they would in the sea. Finally, they popped in a dwarf seahorse raised at the University of Texas's fisheries and mariculture lab.

After turning out the lights, our intrepid trio set out to see what happens when a seahorse sneaks up on a copepod. To do this, they fired a 2.5cm (1 inch) laser beam at the sea creatures and recorded the patterns created when light that has bounced off an object overlaps with light that has passed through it unchanged. The patterns from this technique of 'digital holography' look dull – just jumbled specks of light – but a computer can decipher them to wheedle out where the seahorse was and what it looked like as it ate the crustacean. The beauty of digital holography is that it can take images of the seahorse from any distance, whereas a normal video camera would have the animal going in and out of focus as it swam about.

The slow-food revolution

Dwarf seahorses are top-notch copepod eaters, the laser revealed. A seahorse sneaks up on its dinner so stealthily that on 84 per cent of occasions the copepod doesn't notice and the seahorse ends up within striking range (1mm). Not every successful approach results in a meal. Gemmell remembers one video sequence where the seahorse had sucked up a copepod as normal, except that when it lowered its head the copepod swam out unharmed. 'It literally escaped from the jaws of death,' he says. Generally, though, dwarf seahorses are effective killers. When they're ready to strike, they act fast, sucking up a copepod in less than a millisecond. They catch 94 per cent of the copepods they manage to sneak up on, giving them an overall hit rate of 79 per cent. 'I was astounded at how good the dwarf seahorse is at capturing one of the ocean's greatest escape artists,' Gemmell says. Any copepod next to one that's been scoffed by a seahorse will do its utmost to get away, racing off at up to 36cm (14in) per second.

The team also showed that the all-important 1mm strike region, which lies just above the seahorse's snout, is almost free of turbulence. Further away from the animal, the water is much rougher. The seahorse can thank the shape of its long, pointy head for its ability to create a quiet zone. In other tests, Gemmell found that the three-spined stickleback (*Gasterosteus aculeatus*) – a fish with a blunt head – has no such peaceful water. The dwarf seahorse adjusts its head so that it's perfectly postured for attack: if it swims towards its prey with its head angled too steeply or too shallowly, the strike zone becomes turbulent. But with its head pointing down at 25° to the vertical, the quiet zone above the seahorse's snout is as still as can be. For dwarf seahorses at least, it's the quiet ones that are the most dangerous.

As for solving the problem of turbulence, let's hope someone can put Horace Lamb, up in heaven, out of his mathematical misery. It would be lovely to think that the dwarf seahorse, swimming stealthily around the ocean, holds the secret to one of nature's biggest mysteries – and that it could help someone earn a million dollars.

Air your thoughts

Oranges are not the only fruit and water is not the only fluid. There are also juices, hot chocolates and strawberry milkshakes. And, enveloping the whole planet, air. This invisible fluid is indispensable for life, with about 20 per cent oxygen (the rest is mostly nitrogen and a bit of carbon dioxide). Humans have only recently learned how to access the skies, but some animals evolved to do so long ago, including birds, bats, some insects and a few ancient reptiles.

Animals that fly need to do two things: generate thrust to move forward through the air and create enough lift to counteract the force of gravity. That way, they can fly at a steady height, climb higher if they want to, or launch themselves into the air from a standing start on the ground. This last one's not essential – there are other options if you can't get the lift for it, as we'll hear later. But air is thin; unlike water, it doesn't come close to supporting our weight. Flap your hand about underwater at the swimming pool and you'll feel the water push back. Flap your hand in the air in the changing room and all you'll feel is disapproval. The forces from the air molecules resisting the movement are too small to notice, so generating lift is quite a task. Fortunately, animals can teach us a thing or two about aerodynamics – the study of airflow. Some of today's aerospace engineers are even replicating the bee's approach in tiny drones or micro-air vehicles, or trying to anyway. We're slow learners; we haven't yet got to the bottom of this insect's secrets.

Before we find out how bees fly, and why many 'experts' said they couldn't, a note about flying squirrels, flying frogs, flying lizards, flying snakes and flying fish. Despite their names, these animals can't fly as they're unable to generate thrust to move forward through the air. Instead they jump, then glide. To generate lift and slow their descent, flying squirrels from the Sciuridae family have a membrane between their wrist and ankle so they look like a cloaked Dracula as they swoosh towards the next tree. Flying frogs have enormous webbed feet, flying lizards fold out their ribs along with the connecting membranes to give themselves wings (who needs Red Bull?), while flying fish have large fins. As for flying snakes, they stick out their ribs and suck in their stomachs, making a Frisbee cross-section before launching themselves from a tree and slithering bodies in an S-shape as they would on the ground. But back to those bees.

The bee team

From the outside, it looks idyllic. A life bumbling around flower meadows or buzzing across mistletoe-strewn orchards, rays of sunshine backlighting the black and gold fur spanning your body, as you collect pollen and nectar to take back to your sisters in the hive. They help you care for your mother's larvae in – to humans – an extraordinarily harmonious example of large-scale communal living. The honeybee even does a small dance on its return, shaking its bum in an informational shimmy that tells its hivemates where it found a food-rich patch of flowers (for more on bee communication, see Chapter 6). But the reality isn't so perfect. Bees have a strict role in the hierarchy based on their sex and number of chromosomes, with no hope of changing their job. The luckiest becomes queen bee, but most females are workers, slavishly collecting food, caring for the young and guarding the hive; the few male drones that emerge from eggs laid by the queen in late summer are used solely for their sperm.

Today's bees must also contend with some unpleasant aspects of modern life. Honeybees are under threat from colony collapse disorder, a mysterious syndrome that's devastating hives across Europe and North America. Both the *Varroa* mite and pesticides such as neonicotinoids, which affect bees' memories of where they found food, are under suspicion. And almost all bees face losing their habitat as we destroy orchards, poison flowering 'weeds' with weedkiller, and plough up flower meadows or douse them with fertiliser that makes the grass outgrow the flowers. In their hives, bees are also attacked by mice, hornets and gut-infesting parasites. Another danger are wasps and flies that lay their eggs inside adult bees; when the eggs hatch, the larvae eat the bee from the inside out.

Bees have all that to deal with – and small wings, too. The wings on a worker bumblebee (from the *Bombus* genus) are only 11–17mm (0.4–0.7in) long, about the same length as its body. They're made from a transparent sheet of chitin, the same material as the bee's outer body, or exoskeleton. With these transparent sheets, the bee must heft its body mass of 0.7g (0.03oz) or more, for a worker, into the air. That's barely a sixth of a teaspoon of sugar, which might not sound like much to us but it's a lot for a bee that's keeping itself aloft with just one square centimetre of wing. Making matters worse, the bee sometimes bears its own weight in food as well, in the form of pollen stuck to its hairy legs or nectar in its stomach. So how does the bee do it?

Flight of the bumblebee

Vital as it is for the bee, its flight has proved controversial. Some say scientists in the early twentieth century reckoned bees shouldn't be able to fly because their wings are too small and they don't move fast enough to create sufficient lift. There have even been arguments over who made these claims. Was it French entomologist Antoine Magnan (1881–1938) in his 1934 masterpiece on the flight of insects, assisted by André Sainte-Laguë? Was it physicist Ludwig Prandtl (1875–1953) of Germany's University of Göttingen in the 1930s? Swiss gas dynamicist Jakob Ackeret (1898–1981)? Engineer W. Hoff, using data from entomologist R. Demoll around the time of the First World War? Or even Penry the mild-mannered janitor (*Hong Kong Phooey* (1974), Hanna-Barbera Cartoons)? Or was the claim invented to make scientists of the day, who some felt were getting too big for their boots, look stupid?

What do humans know, anyway? We've said things like 'There is no reason for any individual to have a computer in his home,' 'The Americans have need of the telephone, but we do not. We have plenty of messenger boys,' and 'Television won't be able to hold onto any market it captures after the first six months. People will soon get tired of staring at a plywood box every night.' Fortunately, Charlie Ellington, a zoologist at the University of Cambridge, UK, decided to follow in the footsteps of the bee-perplexed, whoever they were, and solve the mystery. Bees, as he confirmed, are way ahead of us flight-wise: they generate lift in a way we'd never thought of and are only just starting to copy. Even now we don't fully understand how bees fly, despite our powerful computers, jet planes, helicopters and space shuttles. Luckily the bees got the hang of it long ago, producing two or three times more lift than our conventional laws of flight say they should, heaving their bodies into the air, and blithely buzzing about. They're not playing by our rules.

We have lift-off

We're not blaming teachers, but the most common explanation for how things fly isn't completely correct. At school you may have been told that the curves of a plane wing cross-section (known as an aerofoil) cause air flowing over its gently bulging top surface to travel further than air passing beneath. This means, your teacher may have said, that the air over the top must move faster to reunite behind the wing with air that took the lower path. But there's no physical reason why the air that travels over the top of the wing must re-connect with the air molecules that split away and went underneath. And it doesn't; the air takes generally slightly less time to travel over the top of the wing than underneath. But the difference is small and, crucially, the upper air still speeds up, which means the pressure above the wing falls and becomes less than the pressure beneath, sucking the wing up and creating lift. In other words, the force upwards on the wing from air molecules hitting it from beneath is on average larger than the force downwards from molecules above, as there are fewer of them. So the wing is pushed up.

In reality, things are a little more complicated. Which is handy as otherwise planes wouldn't be able to fly upside down. If the above explanation were the whole caboodle, once a plane flipped over, the top surface of the aerofoil would be at the bottom, making air flow faster *beneath* the wing and sucking the plane towards the ground. This would be bad. In the BBC radio sitcom *Cabin Pressure*, pilot Martin Crieff (played by Benedict Cumberbatch) wriggles out of an explanation by distracting air steward Arthur Shappey (John Finnemore) with the news that there's a cat in the baggage hold and he's accidentally switched off the heater. We won't take any measures that drastic, we'll just add that in the real world the aerofoil, whatever its shape, alters airflow a long way off as well as close by, including deflecting air up, then down, then back up again. These movements and the associated accelerations cause forces, and reacting forces in the opposite directions, according to Newton's second and third laws.

You can feel this effect yourself if you stick your hand out of the window of a moving car (after checking for lamp-posts or passing cyclists). By angling your palm into the wind so your thumb is ahead of and higher than your little finger, you'll feel the air forcing your hand, and arm, upwards. Like your hand, a wing doesn't usually face the airflow head on. Instead the front of the wing (the leading edge) is higher than the back (the trailing edge). If you angle a wing correctly it can still deflect the air to produce a lift force even when it's upside down. In addition, the changes in air speed, as the aerofoil moves the air and curves it around its surfaces, also alter the air pressure, as we mentioned above. All these changes – deflection, speed and pressure – feed off each other, setting up a swirl of air from beneath the wing to above. It's complicated, which is probably why your teachers 'simplified' it to the point of error. Let's move on – after all, there are a couple of big differences between a plane and a bee.

Fixed thinking

First, the bee has short wings for its body size. And it doesn't have an engine to create forward motion. Instead, it flaps its wings to create both thrust and lift. This means that conventional aerodynamics, which researchers developed for planes with long, fixed wings, doesn't work for the bee. For its wings to count as fixed, this insect would have to flap slowly compared to its speed of travel, like a big bird. The grey heron (*Ardea cinerea*), which you may see take off from its fishing haunts in rivers, canals, lakes or estuaries as you stare out of a train window, flaps its metre-long wings at a leisurely two beats per second, trailing its long legs behind. Despite this apparent lack of effort, the birds zoom along at 12 metres per second (over 40kph or 25mph). That high ratio of flight speed to flapping rate means we can use conventional techniques to analyse how herons fly. Their long wings suit another assumption of aerodynamic models too – that wings are never-ending. Unfortunately, bee wings are too short and stubby to analyse in this 2D way; their ends have a big impact.

Unlike the 'lazy' heron, bumblebees – or humblebees as they were known in Charles Darwin's day – flap their wings a staggering 150 times a second (another old term for bees and noisy insects, the dumbledore, gave its name to the music-loving headmaster in a series of books about a schoolboy wizard). Mayflies, dragonflies and damselflies have muscles that pull on their wings directly. All other winged insects contract a segment of their body known as the thorax to automatically flap the attached wings. To move its wings fast, the bee has asynchronous flight muscles: each time the bee contracts its thorax muscles they vibrate several times rather than just once. Despite flapping their wings 75 times faster than the heron, bumblebees fly at just a quarter of the bird's speed, at around 3–4 metres per second. The bumbles also rotate their wings between the downstroke (the downwards flap) and the upstroke, flipping them upside down so that the underside of the wing is at the top during the upstroke and the upstroke also generates lift.

So there's no way you can think of bee wings as fixed. Standard aerodynamics says bees should fall from the sky. That's what confused those early twentieth-century scientists. For a fixed wing, the amount of lift depends on the square of its speed through the air (relatively small for a bee), its surface area (again, small), the air pressure (which a bee can't change as it can't control the weather), and the lift coefficient. The lift coefficient depends on the angle of attack – the angle that the wing makes with the oncoming air – and the shape of the wing, which the bee can alter only slowly, by evolving – hardly useful for an insect wanting to take off in a hurry. Put in the numbers and the conventional answer is too low – the lift just isn't enough. But bees fly in the face of convention.

All in a flip-flap

These short, fast-flapping wings posed a bee puzzle (a buzzle?). How could scientists work out how bees fly if they couldn't use their standard aerodynamic tools and model bee wings as fixed and infinitely long? Such a question faced Ellington in 1973 when he arrived in Cambridge from the US, hoping to study how fish swim. Ellington's supervisor, the Danish zoologist Torkel Weis-Fogh (1922–1975), persuaded him to investigate insect flight instead. 'You have to be able to find interesting things in anything you turn your hand to,' Ellington says, philosophically.

Weis-Fogh had just published a paper indicating that quasi-steady aerodynamics (don't panic, we'll get to this in a second) could explain how insects (bees included) fly. The quasi-steady approach is a shortcut to make the calculations easier, if a little less accurate. It ignores the wings' flapping motion by analysing flight as a series of snapshots, where the wing is in a fixed position for each instant in time. The hope is that you can – briefly – use conventional aerodynamics. It looked like everything was solved. Ellington was left to iron out the details by photographing and filming insects as they flew slowly inside glass boxes. This 'near-hovering' flight emphasised the effects of flapping.

To persuade the bees to fly in front of his high-speed camera, Ellington illuminated the centre of each box and blacked out the corners. 'Insects have a mind of their own,' he says. 'As they slowly moved around, you triggered the camera about half a second before they might appear in front of it, and hoped.' After many tries, he'd spend a few days developing the film. Each film was 100 feet long but the camera needed around 50 feet to reach 5,000 frames per second and the last 25 feet thrashed around inside and shredded. 'If you were lucky, there might be 25 feet of film with something on it,' Ellington recalls. 'And if you were very lucky, it might even be a good flight sequence.' With a lot of practice – and wasted film – Ellington and his photographic technician Gordon Runnalls reached a success rate of about one in three.

Unfortunately Weis-Fogh died in 1975. Ellington, missing him greatly, continued the quasi-steady work, studying the flight of insects such as large hawkmoths and buff-tailed bumblebees (*Bombus terrestris*), which he gathered from the Cambridge University Botanic Garden – a now 40-acre haven relocated from the city centre in 1831 by Darwin's botany professor John Stevens Henslow. But the bee wasn't going to divulge its secrets that easily. Ellington faced 'a period of studying insect after insect where this quasi-steady explanation failed [which] pulled the rug out from under us'. History had repeated itself; the latest science again showed that a variety of insects shouldn't be able to get off the ground. In 1984, after over a decade's work, Ellington published 'a big monograph showing that we really don't know how they fly'.

What Ellington needed was a full study of bumblebees flying at a range of speeds, not just near-hovering. So together with Robert Dudley, he developed a small wind tunnel for insect flight that, like those for testing models of aircraft wings, blew a steady stream of air at a speed of the operator's choice. Again, Ellington had to coax the insects into flying in a confined space, this time facing a wind of up to 5 metres per second. 'If they put one wing wrong, they wound up against the downstream barrier,' he says. 'It was difficult, to put it mildly, but the strong visual sense of all flying insects could again trick them.' The team even shone an ultraviolet light from above the tunnel to mimic sunlight and stimulate the bees to fly. As the insects at first flew forwards, backwards or even sideways, the researchers added a striped rotating drum to tell them the direction the world was 'moving', and they duly lined up with it. From these studies, Ellington and Dudley concluded in 1990 that conventional aerodynamics can't explain bee flight at any speed.

Back to square one. The bee was still outwitting us, wind tunnels or no. 'That was a very strange phase,' says Ellington. 'What was causing the wings to generate more lift than they should? We had very little idea, I'm afraid. Somehow our mental picture of it was incorrect.'

To bee or not to bee

With bee flight so tricky, you can hardly blame Ellington for turning from the bumblebee, which has wings around 13mm long buzzing at 150 times per second, to a large North American hawkmoth (*Manduca sexta*, also known as the tobacco hornworm and goliath worm). This insect flaps its 5cm (2in)-long, marbled brown wings a lazy 25 times per second, making it much easier to study. With a brown–and–yellow chequerboard body, this hawkmoth is a beautiful, gentle soul that feeds on flower nectar. It is, though, disturbingly large and reminiscent of the creepy brown posters for serial-killer movie *Silence of the Lambs* featuring a massive Death's head hawkmoth (which, if you use your imagination, has a skull-shaped pattern on its thorax and, unusually for a moth, can squeak). Not being a movie star, the humble *Manduca sexta* has a weirder claim to fame: its larvae keep predators at bay by 'spitting' out nicotine they've absorbed from tobacco plants through their spiracles (air-holes). It's called 'toxic halitosis'. Yuk.

Despite the disgusting habits of its young, this hawkmoth – with wings four times larger than a bumblebee's – is a boon for aerodynamicists. But this quadrupling still wasn't enough to reveal the insect's secrets. So Ellington upped the stakes – he went 10 times larger still by building his own bionic insect, which also meant he could add something that real insects don't do. 'We decided to build a 1-metre-wingspan robot insect that flapped its wings whilst we blew smoke around so we could see what the wings were doing to the air,' Ellington explains. Modelled on a hawkmoth, the robotic 'flapper' ejected vaporised oil from cannily positioned holes along the front of its wings. And what happened to that smoke was a complete surprise.

'If you release smoke along the leading edge of the wing, the one that's cutting through the air, you expect it just to go back across the wing and depart at the trailing edge,' says Ellington. Instead, 'it left the wing, turned 90° and shot off towards the wingtip'. The smoke was doing in-flight aerobatics, swirling around and heading sideways, at right angles to the direction of flight. Aerodynamicists had never seen this before. What was going on?

Smoke signals

The clue lay in research by Weis-Fogh, who had found a way some insects increase their lift over and above convention. He realised the parasitic wasp *Encarsia formosa* and the greenhouse whitefly *Trialeurodes vaporariorum*, as well as thrips, fruit flies, and some butterflies and moths, flap their wings so far up that the tips clap together above their body, like fans dancing to 'YMCA' at a Village People concert. The insects then fling their wings open, forming swirls of air that generate around 25 per cent more lift. 'Clap and fling' is tough on wings, causing wear and tear, and generally only smaller insects use it. Perhaps the swirls in the robot smoke were the key that would unlock the mystery for other insects too?

Like some kind of magic trick, the explanation lay hidden in the smoke. As air hits the front wing of a bee, a hawkmoth or a robot moth, its flapping motion, steep angle and sharp edge create a swirl of air, known as a vortex, above the wing. 'This flow comes around the leading edge, swirls up into a circle and then spirals out towards the wingtip,' Ellington says. Earlier we saw how vortices in their watery form help the pond skater to leg it – they're regions of low pressure where a fluid is moving fast. A vortex above the wing of bees, hawkmoths and robots swirls the air round its centre, sucking the wing upwards and generating extra lift.

Vortices trail back from the tips of wings too, whether they're fixed or flapping. Pilots flying small aircraft must therefore avoid taking off or travelling too close behind large planes to prevent turbulent air in the vortices from the jumbo's wingtips sending the smaller plane crashing to the ground. Vortices are also why geese, pelicans, flamingos and other large birds fly in a V-formation when migrating over long distances. By flying a little behind and to the side of the bird in front, where its wingtip vortices make the air rise, a goose gains extra lift. According to biologists' theories, the best gap is one-quarter of a wingspan, but the birds don't use our rules. Hopefully one day we'll work out why.

Virtuous spiral

So vortices were promising candidates for giving bee flight a lift. But there was a snag. In 2D models of bee flight, which make the sums easier by assuming bee wings stretch on for ever and don't have ends, vortices start small, grow until they're half the width of the wing, then break away. 'We thought that might be one of the mechanisms [the insect] used, but the trouble is the vortex won't stay stable for very long in the models,' says Ellington. 'By the time the wing moves about three or four chord lengths [wing widths], the growing vortex breaks away and you're left with worse-than-normal lift.' Since a fast-flying bee moves roughly eight times its chord length during just one downwards flap, the average lift would be even less than if a vortex *didn't* form.

This breaking-away of the vortex from the wing is known as stall. Like the stall of conventional plane wings, where the airflow also breaks away, it reduces lift. Stall occurs when a wing is too far from the horizontal, making the angle of attack between it and the oncoming airflow too high. On a real plane, stall is bad news and pilots take measures to avoid it. You can also feel its effects if you try to make a remote-controlled model plane soar away by raising its nose. To start with, as you increase the model's angle of attack, the lift increases too. But beyond a critical angle – typically 15–20° for an aircraft wing – the air flowing above the wing separates from its surface and the lift drops sharply, as does the plane.

In the 2D bee-wing models, the three or four chord lengths of travel before the wing stalls is a very dynamic time, giving more lift than expected for steady conditions, so it's called 'dynamic (or delayed) stall'. But what if stall could be delayed longer than usual? Or even for ever?

The bee's knees

Insects are clever. In 3D real life, Ellington discovered, they nip the process of stall in the bud so it never reaches the stall-followed-by-dropping-out-of-the-sky part. They do this by simply stopping the vortex that forms over each wing from breaking away and taking its extra lift with it. We say simply, but it can't be that straightforward because we still don't know exactly how they do it. Somehow, as the robot's smoke trail revealed, insects divert each vortex from spiralling backwards towards their rear, instead forcing it to do a 90° swerve and corkscrew towards the tip of the wing. This sets up a 'spiral leading-edge vortex' and takes away some of the energy.

'If you can somehow suck that vorticity out of it, so it doesn't get too big, then it will stay there,' says Ellington. 'With this trick you can carry on almost indefinitely.' The vortex still grows a little larger as it flows, creating a spiral of ever-increasing girth, like a cone of rolled-up paper. But the vortex spiral doesn't detach from the insect wing until it's about three-quarters of the way to the end. At this point, it joins up with the vortex made by the wingtip to leave an almost ring-shaped vortex in the insect's wake for each downwards flap of each wing.

By making spiral leading-edge vortices and keeping them exactly where it wants them, the insect delays stall long enough to provide up to 70 per cent of its lift, Ellington realised. The rest comes from conventional aerodynamics and maybe from special effects as the wings flip over at the end of each stroke. '[Our discovery] really caught the public fancy as it was very visual as well as being a lovely explanation,' Ellington says. 'In most aerodynamics, the 3D flow field is just a small modification of the field you get in 2D, with flow going front to back over an object. This is one where the third dimension – the flow along the wingspan towards the wingtip – is as large as the flow from front to back. It's a huge aerodynamic effect that hadn't been known before, hadn't been known for propellers even. A spiral leading-edge vortex was a new aerodynamic mechanism.'

So the bee outwits our equations. It stops the vortex created above each wing as it flaps from breaking away, by sending it spiralling towards the wingtip. We're not sure how the bee achieves that direction-change. But keeping the vortex in place next to the wing gives the bee the lift to fly.

Once Ellington and his team knew where to look, they found that rotating propeller blades, like those in helicopters and wind turbines, can create spiral leading-edge vortices too. We now know that other flying animals, such as bats and birds, sometimes exploit leading-edge vortices as well. Some species of bat, for example, use them when flying slowly. But we don't yet know everything. Scientists are still investigating exactly how insects keep their spiral leading-edge vortices from growing too large and how many form on each wing – it could be more than one. And in 2001 Michael Dickinson, then at the University of California, Berkeley, US, found that his model insect formed a leading-edge vortex that didn't spiral in the direction of the wingtip. For Dickinson's model, stopping the flow of air towards the end of the wing didn't, as you might expect, make the leading-edge vortex grow until it detached. Instead, Dickinson reckons, the downwards flow caused by the insect's wingtip vortices can also limit the size of the leading-edge vortex. So the hawkmoth robot smoke trail unlocked only some of the mysteries of insect flight. The bee is hanging on to the rest – for now.

The f(l)ight of the fossil-hunters

The bee and its cunning vortex manoeuvres taught us some new aerodynamics tricks. Now it's time to see what physics can teach us about an animal, not the other way round. Let's turn to pterosaurs – flying reptiles that are long gone but still cause arguments today. That's OK because fossil-hunters, or palaeontologists as they prefer to be called, love arguing over ancient bones. How old are they? What species are they from? Which bone joins onto which bone?

Pterosaurs – from the Greek for 'winged lizards' – lived from 220 million to 65 million years ago. You may know some of them as pterodactyls, the reptiles depicted with varying degrees of anatomical accuracy gliding or flapping about in the background of movies such as the 1925 version of Arthur Conan Doyle's novel *The Lost World* or *King Kong* (1933), as well as *One Million Years B.C.* (1966) and *Jurassic Park III* (2001). These airborne reptiles left no relations but some of them did leave us their bones to argue over. Pterosaur skeletons, first discovered in 1784 by Italian naturalist Cosimo Alessandro Collini (1727–1806), caused confusion even then – was it a fish or a bird? In 1824 palaeontologist Georges Cuvier (1769–1832) reckoned that one specimen resembled 'the product of a sick imagination rather than the ordinary forces of nature'. So what did pterosaurs look like?

Skeletons tell us that the later pterosaurs had forelimbs much longer than their hindlimbs, as well as a short tail and a long thin neck culminating in an almost bird-like head.

The reptiles were probably scaly but some may have had a little fur. Large flaps of skin stretched like a sail between their hind legs and the tips of the fourth (and final) fingers on their front legs. These unclawed fingers were almost as long as the front legs themselves, whereas the first three fingers were short and had claws. Today's bats have similar membranous wings attached to their front legs but they support them with several extended fingers, rather than just one, making the wing more stable. Pterosaurs also had a flap of skin stretching between their hind legs and tail, and a membrane, known as the propatagium, stretching forward from their forelimbs on each side, like a narrow awning.

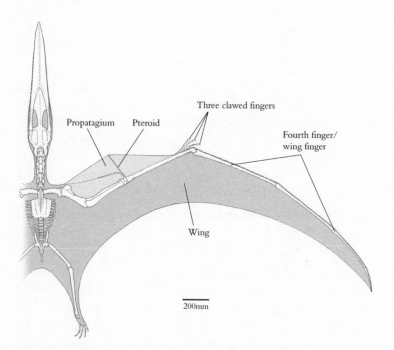

Figure 3.1 **Ptero-soar**. *Pterosaurs like this* Anhanguera santanae *had large flaps of skin stretched like a sail between their hind legs and the tips of the fourth (and final) fingers on their front legs.*

Them bones them bones

Appropriately enough, the bone causing ructions among pterosaur enthusiasts is the one named after these reptiles: the pteroid – a modified wrist bone that's unique to this group. In one pterosaur species, the 4.5m (15ft) wingspan *Coloborhynchus robustus*, the pteroid was around 15cm (6in) long and slightly curved. As wrist bones go, that's big. It's clear that the pteroid supported the forewing. But which way did it point and which bone did it connect to? That question might not sound important to anyone except palaeontologists, but it could have made a big difference to the pterosaur, as Matt Wilkinson of the University of Cambridge, UK, revealed.

When Wilkinson started his work, many experts believed that the pteroid pointed towards the body, roughly parallel to the length of the pterosaur's wing. Other pterosaur cognoscenti reckoned the pteroid pointed forwards, at right angles to the reptile's front limb, creating a much broader forewing.

Pterosaur researchers had almost literally taken sides – some thought the pteroid pointed forwards, some thought it lay sideways. To solve the argument, they looked at more bones. But this didn't prove anything. Over millions of years, the bending, stretching and squashing of the sediments enveloping bones tends to jumble them up. Unusually for people involved in a fossil argument, Wilkinson, who'd worked alongside bee-flight investigator Charlie Ellington, decided to use physics to strengthen his case. 'I realised that the stringent demands of flight would give me a useful way of constraining possible [set-ups] and behaviours in the absence of more definite fossil evidence,' Wilkinson says. 'Using aerodynamics to test the two pteroid reconstructions was the logical next step.'

Time to ptero–soar

Pterosaurs may have inspired the imaginations of modern-day film-makers, but some of them were on the edge, flight-wise. The pterosaur group contains probably the largest animal ever to take to the air. One of the biggest species, *Quetzalcoatlus northropi*, could have weighed 200–250kg (440–550lb), about as heavy as a male lion, even though – like today's birds – it tried to save weight by having hollow bones. But there's a problem with being that heavy. An animal's mass depends largely on its volume and so increases with the cube of its size, whereas its wing area rises only with the square of its size. So if you scale an animal up, its wings become smaller relative to its body weight. As we mentioned when discussing bees, lift is related to wing area, as well as factors such as air speed, air density and the lift coefficient of the wing. To generate enough lift for their heavy bodies to stay in the air, larger animals must have big wings for their size and fly faster.

And so it proved for that monster of the skies, *Quetzalcoatlus northropi*. It had a 12m (40ft) wingspan, which is about one and a half times the length of a London double-decker bus (specifically the iconic 1950s Routemaster model, if you're a transport nerd; today's New Routemasters are shorter). 'Pterosaurs are about as big as a flying animal could possibly be,' says Wilkinson. 'Giant animals have to have proportionally larger wings. Eventually these things wouldn't be able to hold their wings outstretched. Or the minimum airspeed would be so high that when it lands it's going to smash to pieces or it's not going to be able to get enough airspeed to take off, although jumping off a cliff will help.'

Scientists reckon large pterosaurs flew at up to 70kph (43mph). These big animals didn't, however, flap their wings as earlier, smaller pterosaurs – some with wingspans of just 25cm (10in) – would have done. For a *Quetzalcoatlus northropi,* moving its huge wings up and down would have needed a lot of muscle power. Instead, these reptiles probably soared through the air like today's albatrosses, which have wings that are 3m (10ft) from tip to tip. There is, though, one advantage of large, slow-flapping or gliding wings. We can assess pterosaur flight (unlike that of bees)

Above: It's annoying when a wet dog shakes itself dry right next to you, but a mutt's no fool: it could suffer from hypothermia if it stayed still and waited for the water to evaporate.

Above: Roman Stocker's cat Cutta Cutta, who inspired him to work out how moggies drink.

Above: Dogs are messy drinkers. They ladle liquid into their mouths, rather than lap it up like cats.

Above: A male peacock struts its stuff as it tries to woo a peahen, but looks aren't everything: the bird also seduces the female with sounds that humans can't hear.

Right: Barbastelle bats send out extra-quiet ultrasound calls so they can find and sneak up on their moth victims before they hear their attacker and escape.

Below: The California spiny lobster 'plays the violin' to create sound, a technique that works even when it's moulting and has soft surfaces.

Above: Elephants use seismic waves to track other elephants and can even detect the mood of approaching elephants through the rhythm of their footfall.

Right: A Horsfield's hawk-cuckoo chick has 'gape-coloured' patches on its wings, fooling its foster parents into thinking more chicks are present, so they bring extra food.

Right: Most hornets fly from their nest in the early morning, but the Oriental hornet loves the midday sun. The yellow part of the hornet's cuticle allows the animal to absorb sunlight and convert it into electrical current.

Left: The Japanese giant hornet is a dangerous insect, but Japanese honeybees are able to defend themselves by warming their bodies to a temperature that's too hot for the hornet to handle.

Right: Mosquitoes transmit diseases that kill humans, but these insects have their own problems when they suck your hot blood. They cool off by expelling some of it in a droplet.

Right: When a raindrop collides with a mosquito, it's like a 100kg human being hit by a 10-tonne truck. The bug avoids death by moving with the drop and exploiting Newton's second law.

Right: Pond skaters can walk on water thanks to its elastic, trampoline-like surface.

Above: The tokay gecko uses fleshy folds on its feet that are covered in tiny hairs to walk upside down on the ceiling.

Below: Miguel Wattson is the name given to this electric eel at the Tennessee Aquarium in Chattanooga, US, which tweets every time it emits an electric pulse above a certain strength.

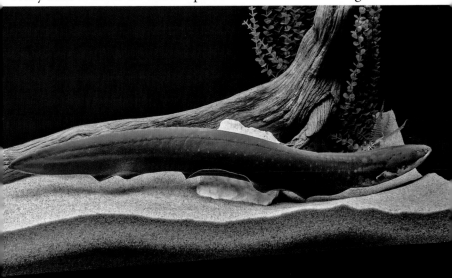

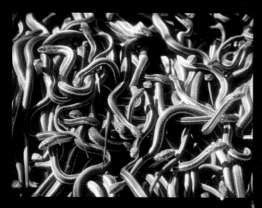

Left: After nine months underground sheltering from the cold Canadian winter, red-sided garter snakes emerge in spring to hang out and warm up in giant, writhing groups hundreds or thousands strong.

Right: An archerfish fires a jet of water from its mouth to dislodge prey on a leaf – a skill that forces it to take into account how light bends, or 'refracts'.

Above: A female loggerhead turtle nesting on a beach in Florida. Loggerheads follow the Earth's magnetic field to find their way back to, or near to, the beach where they hatched.

Right: Ken Lohmann found that hatchling loggerhead turtles swim in a specific direction when exposed to a magnetic field. This turtle is sporting a bathing suit in 'Carolina Blue', one of the colours of the University of North Carolina at Chapel Hill.

Right: The Komodo dragon has a light skull and a bite about as strong as a cat's, but it uses other parts of its body and a 'can-opener' technique to kill prey up to the size of a water buffalo.

The harlequin mantis shrimp snaps out its elbows with one of the fastest accelerations in nature to bash holes in the shells of the crabs and snails it preys on.

with the conventional, fixed-wing aerodynamics that we use for planes. And that makes life a lot simpler.

To help with the pteroid debate, in 2001 Wilkinson created half-lifesize models of the cross-section of a pterosaur wing from steel rods and nylon, then stuck them in a wind tunnel. Would the pteroid-forwards or pteroid-sideways versions fly best? As the inspiration for his models, Wilkinson used 12 incomplete 3D fossil skeletons of *Anhanguera santanae*, another pterosaur species with a wingspan of 4.5m (15ft). Like our old friend with the 15cm (6in) pteroid *Coloborhynchus robustus*, this reptile came from the Ornithocheiroidea group of pterodactyls, which had long and narrow wings. Discovered at the Santana Formation in north-east Brazil, which was once a shallow inland sea, these pterosaurs were encased in limestone after they died, which protected their skeletons from being distorted. 'We have them in their original 3D condition,' Wilkinson says. 'That's really rare, normally they're all just crushed up.' Patching together bones from these excellent specimens helped Wilkinson reveal a nifty physics trick that could have let pterosaurs fly in the face of their large size. One of the pteroid alignments was clearly a winner.

The right angle?

In the wind tunnel, model wings based on pterosaurs with their pteroid bone pointing forwards, which would have given a forewing around 14cm (5.5in) deep in real life, had a big boing. To be precise, they had a high maximum lift coefficient, which is double the lift force divided by the air density, the square of the relative air velocity and the wing area. As the wing's angle of attack increases, the lift coefficient peaks, declining when the airflow separates from the wing and it stalls. If you have a good lift coefficient, you don't need to fly as fast to generate the same lift – the design of your wings doesn't require this extra boost. 'The higher the lift coefficient the further you can push that [flight] envelope,' says Wilkinson. 'I guess the largest pterosaurs were probably pushing that as much as possible.' The wind-tunnel tests also showed wings with a forward-pointing pteroid had a higher lift-to-drag ratio. Drag is a force that acts against forward motion through the air; it's a little like friction on the ground. So a big lift and small drag would have let a pterosaur soar further.

A sideways-pointing pteroid, on the other hand (or wrist), would have made the forewing narrower, just 6cm (2.3in) deep, and resulted in a lower maximum lift coefficient − 1.5 compared with 2.4 for the 14cm (5.5in) forewing. In essence, a pteroid pointing towards the body would have made flying a lot trickier. Wilkinson reckons this aerodynamic evidence is enough to prove that the pteroid pointed forwards.

What's more, he believes the pterosaur could have angled its forward-projecting pteroid downwards on demand. And this would have curved the forewing. 'As the angle of attack goes up and up and up, the leading edge of the wing can hook round so that the airflow remains attached and the wing doesn't stall,' Wilkinson says. 'It makes sure the front edge of the wing is always pointing into the airflow. Because they can alter the angle of the leading edge of their wing even for quite high angles of attack, they can get high lifts at the sort of angles where most wings would stall.' Even if he's wrong and the pteroid did point towards the body, Wilkinson reckons it could still have been mobile, making a deflectable forewing, albeit a narrower one, that curved into the airflow and delayed stall.

At an angle of attack of 20°, Wilkinson's wind-tunnel work showed, a model wing with a forward-pointing pteroid that had curved the forewing into the airflow had a maximum lift coefficient of 2.4. The wing might have reached a higher maximum lift coefficient if Wilkinson had been able to increase the angle of attack past 20° but, as he explains, 'it was clearly topping out by the time I reached the limit of the wind tunnel − it might have gone a bit higher, but only a little bit'. Most aeroplane wings stall at angles of attack between 15° and 20° and have a maximum lift coefficient of around 1.5. So a pterosaur with a forewing-deflecting forward-pointing pteroid would have generated more lift than we can, even with our big brains and computers.

With a lift coefficient this high, a pterosaur could have taken off by holding out its wings and facing a strong breeze, a feature perhaps crucial for those not living near cliffs they could leap off. Many pterosaur skeletons hail from the seaside, where their owners probably ate marine fish, but others lie far from ocean life and from cliffs. And the animals could well have been bad at running to get up speed before take-off because of the large wing membranes attached to their legs. It would have been like trying to sprint in a ballgown.

Like the bee, the pterosaur came up with a cunning technique for preventing stall, although this time it was one that today's aerospace engineers discovered for themselves, without copying nature. If its pteroid bone pointed forwards, the pterosaur managed fluid flow in a similar way to airliners, which have slats along the leading edge of their wings to delay stall.

All in the wrist

There's just one snag. What about those, admittedly flattened, fossil skeletons that show the pteroid bone pointing towards the body? Wilkinson suggests that could be due to another handy feature of a forward-facing pteroid: once the pteroid was fully pressed down, the pterosaur could have pivoted it in towards its body, like a windscreen-wiper, to furl away its forewings. These skeletons could be showing the pteroid in its furled-wing state. That's what Wilkinson reckons, anyway; he feels the aerodynamic evidence is strong, though other experts disagree. For Wilkinson's theory of a forward-pointing pteroid to be correct, the pteroid must join to, or 'articulate with', another wrist bone called the medial distal carpal. And nobody has found a fossil where these two bones are next to each other. Instead the pteroid bone is always found next to the sesamoid – a small bone that, like the kneecap, increases the pulling power of the muscle it sits within.

Wilkinson has an answer for this too. He thinks the sesamoid could have lain inside the muscle that extended the pteroid. 'After death the tension in the forewing may pull the pteroid out of articulation so that the sesamoid ends up looking like it's articulating here. I don't see anywhere else that the pteroid could possibly articulate.' To get to the bare bones of the matter, the sesamoid may be knuckling its way into the wrong place, between the pteroid and the medial distal carpal, confusing palaeontologists as it goes.

Wilkinson still believes that 'in the fullness of time' we'll show that the pteroid did lie next to the medial distal carpal, proving it pointed forwards. 'Ideally we need more 3D [skeleton] material, and unfortunately these things are very hard to come by,' he says. In the decade since Wilkinson wrote his paper, no better specimens have come along. And the pteroid-forwards/pteroid-sideways how–did–these–big-guys-fly debate won't be over until – as the old song nearly goes – a fossil pterosaur appears with its pteroid bone connected to its medial distal carpal bone. The pteroid, it seems, is still a bone of contention.

Fluids on the brain

This chapter's been a game of two halves: first liquids, then gases. Surface tension, that stretchiness on top of a liquid, stops a pond skater from sinking and the insect walks by shedding hemispherical vortices. Cats drink by touching their tongues onto the surface of a liquid and creating a column of fluid as they lift their tongues into their mouths. And dwarf seahorses in the warm waters of the Caribbean eat by creeping up on their prey, keeping the water as still as possible. Turbulence is a no-no.

On the gases front, no-one knew how bees generate enough lift to counteract the force of gravity until a robotic insect and some smoke revealed many of their secrets. Bees stop the vortex created above each wing as it flaps from breaking away, instead sending it spiralling towards the wingtip. Meanwhile, studying fluids has added to a great fossil debate by revealing that pterosaurs would have flown more easily with a pteroid pointing forwards, not sideways.

Now we've aired our views about gases, we can move on to our next topic. It's a wave that ripples silently and helps animals – if they can detect it – to communicate, find food, stay safe, avoid bumping into things and even reproduce. It's time to sound off about sound.

CHAPTER FOUR

Sound: Good Vibrations

SEDUCTIVE PEACOCKS * THE BATTLE OF THE BATS
* SNAKES WITH KILLER HEARING * TRIANGULATING
ELEPHANTS * LOBSTER VIOLINS

'I videotape peacocks having sex.'

It's what Angela Freeman used to say between 2009 and 2012 when people asked her what she did for a living. A research student at the University of Manitoba, Canada, Freeman hung out at her local zoo with colleague James Hare, filming these beautiful birds. Together the two voyeurs spied on a total of 37 peacocks – the males of the species *Pavo cristatus* – as they strutted around, wooing peahens.

As Freeman's tapes show, the peacocks know exactly what turns the females on. Peahens, which have stumpy brown tails, love it when a male fans out his long tail feathers into a semicircle, like the train of a wedding dress held high by invisible bridesmaids. Each feather sports green rings filled with brown and a deep blue spot at the centre. For the peahens, however, these eye-spots are more than eye candy.

Peahens plump for peacocks with long trains as these guys tend to be fatter, proving they're better at finding food. The females generally – though not always – prefer males with lots of big, bright eye-spots as they give the females chicks that grow faster than youngsters from less attractive dads. When he fans out his tail, the male's showing off what an excellent father he'd make. It's a visual clue that he has good genes.

But looks aren't everything when wooing peahens. That's where Freeman and Hare come in. Watch a peacock strut its stuff in a zoo, a park or the grounds of a stately home and you'll hear a huge racket too. Peacocks coo and caw, creating short, loud blasts like those from cheap blowers at kids' parties – the ones with a plastic mouthpiece and a rolled-up pipe that straightens out when you puff into it. To humans, the peacocks' raucous noises are at odds with the beauty of their plumage. Listen closely, though, and you'll also hear peacocks making a quieter, and more pleasing, shivering noise as they rustle their tails.

Our two animal spies wanted to find out more. What they discovered was a surprise. It turns out peacocks don't just make a rustle and a racket, they also create sounds that we can't hear. So how did Freeman and Hare detect sounds that were inaudible to humans? More importantly, why do peacocks make them in the first place – to attract mates, to deter rivals, or both? Before we get to the bottom of that mystery, let's first look at what sound is.

Sounds good

The poster for the 1979 sci-fi film *Alien* holds a clue. 'In space,' it warns, 'no one can hear you scream.' That's because sound needs something to travel through. Here on Earth that something is usually air, although sounds can also pass through solids, as you'll know to your cost if your neighbours have cranked the music up and bass beats are thumping through the wall. Noises can travel through liquids too – you can test that by blowing bubbles in your bath water. But up in space, which is almost a perfect vacuum, there are no sounds. So next time you watch a Hollywood movie where a spaceship explodes with a loud bang, you'll know the directors haven't done their homework.

To make a sound – any sound, rustle, racket or pop – you need an object moving fast enough to set a medium in motion (we don't mean a psychic but a substance such as a solid, liquid or air). Try humming. You'll have to blow air from your lungs with enough strength to vibrate your vocal cords. The vibration is the key, as with any sound, be it the voice of your favourite singer belting out from a loudspeaker or the twang of a guitar string. The vibrating vocal cords, speaker membrane or metal wire push against neighbouring air molecules, moving them closer together. The molecules in this compressed region rush away towards less crowded air, before crashing into other molecules further out. This new group picks up some of the original molecules' momentum and goes on to push together other molecules even further away.

The result? Sound ripples through the air as a wave of alternating regions of high and low pressure. Like all waves, sound has a frequency, which is the number of ripples, from high pressure to low and back again, that pass a particular point every second. Frequency is a handy measure as it lets us categorise different sounds. Its unit is the hertz (Hz), named in honour of the German physicist Heinrich Hertz (1857–94), who transmitted the first radio waves in 1886. Middle C on a piano is 262Hz. Human voices are a mix of frequencies, but the basic vibration ranges from 80Hz for a guy singing deep bass up to 1,100Hz for a shrieking toddler. The top note on a piano, four octaves above middle C, is even higher at 4,186Hz.

Animals don't play music, but sound is incredibly valuable to them. These waves of high and low pressure help wildlife to find prey, detect danger, communicate with potential mates and explore their surroundings. As we'll discover in this chapter, animals use sound everywhere, from bats flying high above our heads, to snakes slinking beneath the sand, to crustaceans deep underwater. But for a sound to be useful, an animal must hear it. Humans hear when changing air pressure reaches our ears and vibrates our eardrums at the same frequency as the sound wave. The three bones of the middle ear – the hammer, anvil and stirrup – pass this vibration to the inner ear, setting up a wave in the fluid it contains, which in turn wiggles hair cells that transmit nerve signals to the brain. So to answer the conundrum of whether a tree falling in an empty forest creates sound: yes, there is a sound wave, but you have to be there to hear it.

We're going to the zoo

Back to those amorous peacocks and their strange sounds we can't hear. Freeman and Hare didn't have the money to fly to the peacocks' native lands of India, Pakistan and Sri Lanka and watch birds in the wild. Instead they set up shop at Assiniboine Park Zoo in downtown Winnipeg with audio equipment primed to record the noises peacocks make with their trains, including sounds inaudible to us. They also had video cameras to see how close peahens get to a potential mate and, crucially, if any of the pairs had sex.

Watching peacocks in comfortable conditions – no late nights, no mosquitoes, no venomous snakes, and no extreme temperatures – might sound cushy. But the work was 'exceptionally difficult' for Freeman. For a start, if it was windy or raining she couldn't capture any decent audio. Then there was us lot – human visitors, who'd talk in loud voices just as a peacock displayed its train, disturbing the recordings. Staff at Assiniboine helped by letting Freeman and Hare in before opening time so they could bag early-morning data in relative peace, but there was another, unexpected problem: the zoo's wild turkeys. 'The peacocks would perch on the picnic table and the turkeys, who don't perch, would run around the tables in circles trying to figure out how to get at them,' Freeman recalls. It brings a whole new meaning to table scraps.

Despite these snags, Freeman and Hare bagged plenty of peacock sounds. Feeding their recordings into a computer revealed the frequencies the peacocks had been squawking at, and for how long. High-pitched audible sounds from the peacocks' throats generally lasted a second or two and ranged from about 400Hz up to 20,000Hz. But our two researchers also noticed strange signals down around 3–6Hz. Just as we cannot see infrared light, which has a lower frequency than ordinary red light, so we can't hear these low-frequency sound waves, or 'infrasound'. It's not just peacocks: a handful of other animals, including hippos, whales and elephants, generate infrasound too. Even earthquakes, erupting volcanoes, stormy ocean waves and meteorites falling to Earth make such low-frequency sound.

Spooky sensations

We can't hear infrasound as our ears stop functioning below about 20Hz, roughly 9Hz deeper than the lowest note on a piano, which is the A three and a bit octaves below middle C. Some of us might, though, be able to detect low-frequency sound in other ways. In 2003 the British composer Sarah Angliss held an experimental concert at London's Purcell Room, playing music laced with infrasound at a frequency of 17Hz. More than a fifth of the audience claimed to have felt anxious, scared or sorrowful, or to have sensed chills down the spine. It's intriguing to imagine they had somehow sensed the infrasound, though there may have been a simpler explanation – they were just scaredy-cats.

Peacocks aren't looking to frighten, though. It's sex they're after. By matching their video footage to their audio files, Freeman and Hare discovered that the male generates two kinds of infrasound. Both involve the brightly coloured feathers that he fans out to create his train (rather than the short brown feathers beneath). The peacock creates infrasound between 2.8Hz and 4.2Hz by rapidly 'shivering' his long train feathers, so that each one ripples up and down. In the 'pulse train', on the other hand, the peacock vibrates his feathers starting at the end nearest his body and moving out. At 3.1–6.4Hz, the pulse-train sound has a slightly higher frequency overall. But why use two so subtly different infrasounds?

For the answer, we need to sound out the physics more closely. If you live under a flight path, you'll know the relief every time an aircraft disappears into the distance. Each plane's not getting quieter as it flies off – the motors are still working full pelt. It's just that the sound from the engine, which starts off in a small space, spreads in all directions like an ever-inflating beach ball, weakening as it goes. But sound also fades with distance because it takes energy to vibrate the air molecules and overcome their reluctance to move, meaning the sound continuously loses energy to the air. Since high-frequency waves vibrate the air more often in a given time, they lose energy faster than low-frequency waves. That's why you can hear the bass line from the students partying down the end of your street, but not the vocals. Being higher pitched, the voices fizzle out faster.

This 'attenuation' of sound – its weakening with distance – holds the clue to peacocks' 'music' too. As Freeman's tapes showed, the infrasound a peacock makes depends on how far he is from the object of his desires. If the peahen is less than 5m (16ft) away, her suitor mostly uses pulse trains. But if an attractive peahen is further off, the male mainly creates infrasound by shivering his feathers. The sound from a shiver has a lower average frequency than the noise from a pulse train so it attenuates less and travels further, making him more likely to attract the attention of distant females. After all, there's no point whispering the peacock equivalent of sweet nothings at a peahen who's too far off to hear.

But how can we be sure these birds pick up infrasound? Their romantic mission not yet complete, Freeman and Hare played pre-recorded peacock soundtracks to the birds at Assiniboine through loudspeakers. Both males and females became more alert when they heard a mix of audible sound and infrasound from a feather train, our peafowl researchers found. The peahens ran around, while the peacocks called out and faced the loudspeaker unit. When speakers pumped out only the trains' audible signals, on the other hand, neither females nor males responded. Infrasound must be a key part of how the birds communicate: males called out even when infrasound alone was played. What's more, peacocks who'd carved out an area of land as their own – territory holders – were more responsive to infrasound than males who'd simply wandered past Freeman and Hare's loudspeaker unit. Case closed. Peacocks create infrasound to attract peahens and detect it to protect their territory from competitors.

For our understanding of attenuation we can thank the physicist George Gabriel Stokes of Navier-Stokes equation fame, who first studied it at the University of Cambridge, UK, in the 1840s. Animals were far from Stokes's mind when doing his pioneering work, but he did have a brush with peacocks later in life. Sometime in the 1860s, Stokes received a letter from Charles Darwin, who raised a question first asked by Isaac Newton in the early 1700s. Newton had wondered if the green and blue colours of a peacock's eye-spots were due not to pigments in the feathers, but to their microscopic structure making light bounce off them in unusual ways (see Chapter 6).

'Will you have the kindness to attend to one point,' Darwin asked Stokes, 'namely, whether a gradual thickening or thinning by little steps from the centre to the circumference, of the film of colouring matter wd account for the zones of colour which occur; or must there be zones of different kinds of colouring matter?' With some simple experiments, Stokes confirmed both Newton's and Darwin's suspicions. He could never have dreamed, though, that his studies of the attenuation of sound would, more than 150 years later, solve the mystery of why peacocks so carefully tailor their sound.

How these birds detect infrasound remains a mystery, but it's clear that a peacock's hearing is better than ours at low frequencies. That's not to say we're great at picking up high frequencies either. Most of us hear best in the middle-of-the-road 2,000–5,000Hz mark – roughly the top couple of octaves on a piano keyboard. Our hearing peters out at higher frequencies, especially as we get older or listen to too much loud music (just ask any faded rock star with bad hearing). Even those of us with perfect hearing can't detect anything much above 20,000Hz. This 'ultrasound' is inaudible, which is why a pregnancy scan is silent apart from the yelps of joy when parents first 'see' their fetus. Generated by electric currents vibrating certain crystals, the ultrasound in a hospital scanner mostly sails through the mother's belly but bounces back out if it hits something hard. The result: a much-cherished, if slightly fuzzy, image of the baby – white for reflective bones, black for fluids that let the sound continue on its way and grey for squishy tissue.

While people can't hear ultrasound, some animals can. A dog whistle produces ultrasonic waves between 23,000Hz and 54,000Hz, perfect for bringing your mutt to heel without annoying other dog walkers. Cats hear ultrasound too, as can dolphins and porpoises. But there's one animal that's king of high frequencies. It's time to leave the ground and look up to an animal that flies.

Out to bat

Night-time in the parklands of an English stately home. A giant London plane tree towers above the sweeping lawns and the moonlight reflects silver from a river. It's the kind of place you'd expect to see peacocks, or at the very least a ghost. The evening is mild and the air fresh with the scent of leaves and night-blooming flowers; moths are out in force too. From time to time the fluttering silhouette of a bat breaks the indigo of the sky. It twists and turns, vacuuming up insects.

But it's dark. How can a fast-flying bat grab a speeding insect on the wing? It's a feat more impressive than any circus trapeze act, where at least the grabbee wants to be grabbed. This nocturnal activity could be a hangover from a time 220 million years ago when bats' mouse-like ancestors evolved from reptiles and wanted to avoid being eaten. Living in the dark meant they could keep clear of daytime-active dinosaurs. On the downside, it was hard to see where they were or where they were going. When they first took to the air, some of these early mammals developed a primitive system of firing pulses of ultrasound to learn more about their surroundings.

Not that these early bats knew what they were doing. 'It probably started with a very basic analysis of echoes coming back, using sounds originally made for a different purpose, such as communication,' reckons Holger Goerlitz, a bat expert at the Max Planck Institute for Ornithology in Bavaria, Germany. 'It's like when you go into a church or a cellar and you talk or make sounds with your feet, you can already hear that it sounds different, that it's a big or small room.'

Bats create high-frequency pulses of ultrasound just as we talk and sing, vibrating their vocal cords and then 'shouting' the sound out through their mouth, or, for some species, their nose. By noting how long it takes the ultrasound to bounce back to them off trees, cave walls and other reflective surfaces, bats can suss out how far away these objects are. A wave that returns fast must have hit something nearby, while one that gets back later must have travelled further. Thanks to this 'echolocation', bats can navigate their way past obstacles.

Ultrasonic bat calls are among the loudest sounds in the animal kingdom. To avoid deafening themselves with their own noise, some bats contract muscles in their ears while they shout, to close them up. Others emit sound at a frequency they can't hear – at least, not until the noise bounces back. For this nifty ear-protection, bats exploit the physics that sees the pitch of an ambulance siren rise as it approaches but fall as the vehicle races away. Known as the Doppler shift after Austrian physicist Christian Doppler (1803–53), the frequency goes up as the ambulance drives nearer because more sound waves must pack into a smaller distance. The frequency drops as the vehicle moves off because the waves are stretched out. The Doppler shift applies even if the ambulance isn't moving but you are; it's the relative motion between you and the vehicle that counts. So when a bat that can't hear its own ultrasound flies towards a returning sound wave, the echo is Doppler shifted to a higher frequency that now lies in the range that the bat *can* hear. Crucially, the echo has lost energy on its round trip and is quieter than the original ultrasound pulse, leaving the bat's ears unharmed.

Many bat species not only get from A to B without crashing into large objects but can even echolocate tiny, fast-moving insects, such as moths, for supper. That's because a moth's beating wings also Doppler shift the bat's ultrasound pulses, as well as reflecting back different amounts of sound towards the bat as they flap. The result: a sound 'signature' the bat can look for in its echoes, helping to track the insect down.

Ear we go

Despite their impressive echolocation powers, bats don't have everything their own way. Many moths have developed ears that hear ultrasound to warn of hunting bats. Fighting back works: moths that have ears tend to be preyed on less by bats than their non-eared mothy cousins. So is it worth bats' while to retaliate and come up with a counter-counter-measure? Missing out on a moth kill is inconvenient for a bat but it'll live to fight another day, just one meal down. The moth, though, has it all to play for. If it loses the battle, it'll be eaten. Biologists call this skewing of the odds the life/dinner principle.

Keen to see if bats have upped their arms-race ante, Goerlitz and colleague Matt Zeale examined droppings from a group of 51 barbastelles (*Barbastella barbastellus*) to see what they'd eaten. Named after the Latin for 'star beard' because of the white hairs that grow from their top lip, barbastelles' ultrasound is loudest at 33,000Hz, well within the hearing range of most eared moths. A moth with any sense, you'd think, would hear the bats' ultrasonic beeps and fly away. But Goerlitz and Zeale got a surprise. Barbastelles, according to the genes in undigested bits of their dinner, feed almost entirely on moths *with* ears. This bat, in other words, relishes eating the very moths that are listening out for it. What, the researchers wondered, is going on?

I just called...

To find out, Goerlitz travelled to Mottisfont, an eighteenth-century mansion north of Southampton in the UK. Built on the site of a former priory and set among extensive gardens, Mottisfont is one of the few sites in Britain where barbastelles breed. Goerlitz is an enthusiast for batty field work, despite it being dark, tiring and hard to see unless it's early evening. 'Some of the guys come out whilst it's still kind of dim,' he says. 'I'm fascinated when you can see them quickly flying by a hedgerow or against the sky. You can see how they suddenly dash for an insect – they fly along and suddenly they make a turn and catch something.'

Goerlitz listened to the bats using a detector that converts ultrasound calls to frequencies audible by humans. 'You hear their regular calls, tchk, tchk, tchk,' he says, 'and then ZCHK they produce a very fast sequence when they make a catch. You get a feeling for what they're doing by just listening to them.' In a second surprise, the barbastelles' ultrasound waves were 10–100 times fainter than Goerlitz expected. This ultrasonic whisper means that barbastelles can only detect insects that are less than 5m (16ft) away. Any further and the bat's returning echoes are too faint for it to hear. Yet louder-shouting species of bat can echolocate prey from up to 15m (50ft).

Attenuate, attenuate

Let's take stock. First, the barbastelle prefers to eat moths that hear it coming, rather than moths that *can't* hear. Second, with its quiet call the bat can only find eared moths that are right up close. But this strategy is so effective that biologists have even coined a name for it: 'stealth echolocation'. So how does it work?

After all, physics is not on the bat's side. The echo bouncing back to the animal is quieter than the sound pulse reaching the moth. That's because it's travelled double the distance – from bat to moth and back again – so it loses twice the power through attenuation, the same petering out we met with peacocks. Worse still, the bat's ultrasound signal weakens further when it reflects off the moth. This double whammy means that, all else being equal, it's harder for the bat to detect the moth than it is for the moth to hear the bat. The bat will need to be closer to what it's listening out for than the moth needs to be. Does this mean the laws of physics are conspiring to disadvantage the bat?

Goerlitz and his colleague Hannah ter Hofstede decided to switch their attention from the barbastelle to its moth prey, focusing on the large yellow underwing (*Noctua pronuba*), one of the bat's favourite meals. With a wingspan of around 50mm (2in), this moth has orange hindwings that peek from beneath its mottled grey-brown forewings when it spreads them apart or flies. And it's got ultrasonic ears too. But, the researchers found, large yellow underwings don't hear nearly as acutely as the barbastelle.

So how does this discrepancy in hearing power affect the barbastelle/moth stand-off? The conflict between the physics working against the bat and its good hearing means there's a 'sweet spot' for ultrasound at a particular noise level where the bat's just near enough to echolocate the moth and the moth's just near enough to hear the bat. At this position the bat's better hearing exactly compensates for the echo it receives being weaker (because of the extra attenuation) than the sound going directly to the moth. Bat 0, Moth 0. If, however, the bat beams out ultrasound louder than this noise level, it's a 'moth win' as the moth hears the bat from further away, whilst the bat can't detect the echo, and the insect flies away. Bat 0, Moth 1. Only by beaming out ultrasound more faintly than this threshold can the sharp-eared barbastelle detect the eared moth without the moth hearing it. Bat 1, Moth 0. The ultrasound pulse from

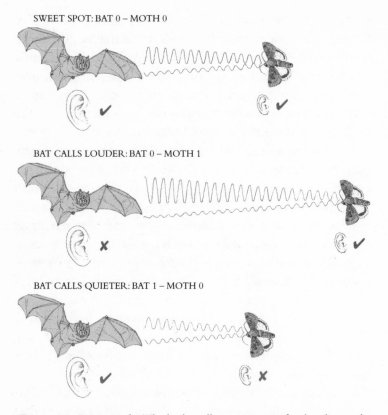

SWEET SPOT: BAT 0 – MOTH 0

BAT CALLS LOUDER: BAT 0 – MOTH 1

BAT CALLS QUIETER: BAT 1 – MOTH 0

*Figure 4.1 **Bat attack**. The barbastelle compensates for the physics that attenuates its ultrasound calls on their journey to and from its prey with extra-good hearing.*

the bat is quiet and the echo that bounces back is quieter still, but the bat senses it because its hearing is so good. The sound reaching the moth is louder, as it has only travelled one way. But, being a cloth-ears, the moth can't detect it. Unlucky, eared moth: you're dinner.

It's as if the bat is stalking the moth, creeping up so quietly that the quarry doesn't hear its hunter until too late. A typical large yellow underwing hears a barbastelle only when it's flown to within 3.5m (11.5ft), Goerlitz learned; any further away and the bat's call is too quiet. Given that a barbastelle flies at about 8 metres per second – a shade slower than an Olympic sprinter runs – it covers those 3.5m in less than half a second. Once it's heard the bat, the moth has little time to escape. Taking into account reaction times and the fact that the insect waits for another, less sensitive, auditory nerve to trigger before flying in erratic zigzags to evade capture, that half-second is not long enough. For the moth, it's as if the bat appeared from nowhere.

The barbastelle, at least, has taken the next step in the battle, emitting ultrasound quietly so it can reach its prey undetected. The obvious move for the moth would be to improve its hearing. Large yellow underwings haven't done this, Goerlitz believes, because startling at every single sound they hear – such as crickets chirping or bats too distant to be a danger – would be a waste of time and energy. There's no point reacting to threats that aren't real.

But why has the barbastelle overcome its attenuation handicap to chase eared moths when other moths are available? It's a question of competition. Since no other bats have sussed this physics like the barbastelle, it can gorge on as many eared moths as it likes, leaving other bats to contend for moths without ears. With its tactical ultrasound nous – good hearing and extra-quiet echolocation – the barbastelle is a winner. There's nothing batty about this bat.

Snaking through the sand

A cobra lies coiled in a basket on a dusty street in a village in southern India. Grey-brown and about a metre (3ft) long, this venomous snake hides under a straw lid. Just out of harm's way is an Irula tribesman wearing baggy orange trousers, sitting cross-legged on a mat. The turbanned figure blows into a *pungi*, a wooden flute that is like a recorder with a bulbous middle. As a crowd looks on, haunting sounds blare out whilst the man plays his fingers over the holes of the *pungi*. The snake rises from the basket, its upper body vertical, with its head swaying from side to side as if mesmerised by the music. The tribesman seems to have the snake under his control.

This party trick has kept snake charmers in business for centuries. But there's nothing psychic about it. Snakes don't have ears and can't hear the *pungi*. The reptile instead watches the instrument's movement, tracking it like it's a dangerous enemy. It's not a charming activity either: owning snakes has been illegal in India since the 1970s.

Lacking lugholes, snakes are also the last animal you'd expect to find in a chapter on sound. But there's more to snake hearing than you might think, as two US biologists, Bruce Young and Malinda Morain, discovered when playing snake tricks of their own. After working with 'two very large king cobra that would growl like angry dogs' whenever he approached them to study the physics behind those noises, Young became interested in 'the other half of acoustics' – how such animals perceive sounds.

Young and Morain bought four venomous Saharan sand vipers (*Cerastes cerastes*). About half a metre (20in) long, with a horn of skin above each eye, these yellow-brown snakes live in the deserts of North Africa. Young had kept several sand vipers as pets and loved to watch them hunt. 'They lie buried in the sand and then suddenly erupt, up-turn in mid-air and strike a mouse,' he says. They're masters of ambush, waiting coiled with only their head peeking out, for rodents or lizards to scurry past.

After watching his pets, Young wondered how the Saharan sand viper senses its prey. Since the snake's eyes are under the sand, the reptile must locate its meal not by looking for it but from other cues. Being underground, it can't find mice by flicking its tongue to 'taste' chemicals in the air, as many snakes do (it would just get sand in its mouth). To get to the bottom of things, Young and Morain placed their four snakes on a bed of sand in a 2m by 2m (6.5ft by 6.5ft) glass terrarium at Lafayette College in

Pennsylvania. 'Saharan sand vipers are very good research animals,' says Young. 'My sandbox had high enough walls that the snakes couldn't escape and my lab was secure so I didn't have to worry about a stranger being bit. I would come in every morning and see the low mounds that are formed when they burrow. It reminded me of my time in India with the Irula tribesmen who are amazing snake trackers.'

With the terrarium kept toasty warm by heat lamps mimicking desert conditions, Young and Morain put a mouse in the cage at the furthest point from the snake, and tracked its progress around the lair with high-speed video cameras. When the mouse neared, the snake struck out fast – biting, holding, then killing its prey. The scientists studied 18 mice, each time noting the rodent's location when the snake attacked.

So far no surprises; the Saharan sand vipers behaved in the lab just as you'd expect in the desert. Now for the researchers' tricks. Young and Morain blindfolded each sand viper with surgical tape so it couldn't see. They also blocked up the snake's nose so it couldn't smell. Back in a cage with a mouse, each Saharan sand viper could strike its target as well as it could before its nose- and eye-job. Unable to see or follow its nose, the Saharan sand viper could nevertheless sense its prey.

Suspecting that the snake picks up the pitter-patter of the mouse scurrying across the sand, Young and Morain created artificial vibrations to see how the snake would react. First they got hold of a Styrofoam ball roughly the size of a mouse, which they placed in their sandbox. The next step was to warm the ball to mouse temperature (37°C) with a heat lamp, before adding a blindfolded and nose-bunged sand viper to the sandbox. With a thin balsa wood stick, the researchers gingerly pushed a ball right up close to the snake's head. Then they tapped the polystyrene to set up vibrations in the ground, taking care not to touch the snake. Immediately, the snake flicked its tongue and struck out towards the ball. Despite being unable to see or smell, the Saharan sand viper had detected the ball's vibrations and deduced that a 'real' animal was scampering by.

A jaw of two halves

Keen to discover how the snake senses vibrations, Young teamed up with two physicists, Paul Friedel and Leo van Hemmen from the Technical University of Munich in Germany. Van Hemmen had become fascinated by snakes after studying hearing in barn owls and scorpions. 'I found it funny that many snakes live in the desert – in sand – and wondered how they hear given they don't have ears,' he recalls. 'It took me a while to discover that they aren't deaf but their hearing ability is totally different from anything else in nature.'

While snakes don't have ear-like holes in their heads, they aren't entirely earless. These reptiles have an inner ear containing a cochlea – the spiral, bony chamber that, in other animals, channels sound waves to cells that detect them and send them to the brain. These structures remain from the 'real' ears snakes had before they evolved for life underground, where an earhole would fill with dirt. Could this ancient structure still be working, despite having lost a key part – the hole to the outside world? If so, how do snakes sense sound waves if their ears are sealed?

It was time to look at another part of the snake's head: the Saharan sand viper's lower jaw, which – like all snakes – is built from two halves. Unlike a human jaw, the two halves aren't joined rigidly. Instead, they're linked by an elastic ligament that lets them spread apart (the upper jaw can't do this trick). Moving each half of its lower jaw almost independently, the snake can force its mouth wide open and cram prey – head first – into a gaping hole. This flexibility is what allows a python or boa constrictor to swallow something as big as a deer, far larger than its mouth.

Van Hemmen noticed that the Saharan sand viper always rests its head on the desert floor. If sound waves from the prey travel through the sand and set the lower jaw vibrating, he reckoned, this motion would pass from each half of the jaw up the 'quadrate' and 'columella' bones at the back of the snake's head and into the inner ear. There, the hairs inside the cochlea would vibrate too, stimulating neurones that send signals to the snake's brain. But wait! Which sound waves? There's something about sound in solids that we've so far brushed under the carpet.

Going underground

Time to come clean: sound travels through solids in several ways, as you'll know if you've experienced an earthquake. Some sound waves move like vibrations in air, rippling the ground backwards and forwards in the same direction as the sound travels. They're 'longitudinal' waves in physics-speak. When created by earthquakes, these 'primary' waves zip through the interior of the Earth at about 5–8kph (3–5mph) yet are relatively mild, merely making your windows rattle. 'Transverse' sound waves, meanwhile, shake a solid up and down perpendicularly to their direction of travel on their journey through our planet. In earthquakes, such 'secondary' waves are more destructive. Moving more slowly, at 3–4kph (about 2mph), they arrive after the primary waves, the time gap depending on your distance from the epicentre of the earthquake.

A little later you'll feel a third kind of sound wave that's travelled through solids, one that's crucial to the Saharan sand viper. Predicted in 1882 by the British physicist Lord Rayleigh (1842–1919) at the University of Cambridge, these waves are a mix of transverse and longitudinal motion. Rayleigh waves from an earthquake travel just below the surface of the Earth (rather than through the

interior) at a sluggish 50–300m (165–990ft) per second.
Once they finally arrive, however, these waves violently
shake the ground up and down *and* from side to side,
sometimes even making buildings collapse (hardly
surprising that sounds in solids are called 'seismic' waves,
from the Greek for 'shaking'). If you stood in your garden
and stared at a point in the ground as a Rayleigh wave
passed from left to right, that point would – due to the mix
of down/up and left/right motion – move anticlockwise
following the outline of an ellipse. The ellipse would be
pointy and narrow because such Rayleigh waves' up-and-
down motion is bigger than their oscillation from left
to right.

Saharan sand vipers don't care about Rayleigh waves
from earthquakes. But when a mouse scampers across the
desert, about 70 per cent of the energy it transfers to the
sand turns into these waves, which travel below the surface
a little slower than a Rayleigh wave from an earthquake, at
about 45m (150ft) per second. They have a frequency of
200–1,000Hz, roughly from middle C on a piano to two
octaves above. Unsurprisingly, as well as being slower, the
Rayleigh waves from a mouse are much gentler than those
from earthquakes, with their amplitude – the size of a
vibration – barely a thousandth of a millimetre.

Feeling those vibrations

A sand viper that hears mice by letting their Rayleigh
waves jiggle its jaw sounds like a smart idea. Van Hemmen
had good reason to believe it was true: tests have shown
that snakes can respond to vibrations as small as
one-hundred-thousandth of a millimetre applied to their
lower jaw. But even if a sand viper can pick up the ground
trembling from a scampering mouse, that's the job only
half done. The snake also needs to pinpoint where its
potential dinner is. Is the animal disturbing the sand back

there behind a bush? Or somewhere over to the left? The snake hunts at night so it can't see. Besides, Young and Morain's blindfold proved that the sand viper doesn't use its eyes to spot prey.

Again, the secret to its hunting lies in the Saharan sand viper's two-halved jaw. Unless the mouse is straight in front or directly behind, its Rayleigh waves strike one side of the snake's jaw before the other. The resulting vibrations reach the inner ear at slightly different times, sending separate signals to the brain. For a mouse walking exactly to the right of the snake's jaw, the waves arrive at an angle of 90° to the length of the jaw and reach the right side of the jaw about half a millisecond before they get to the left. But if the mouse moves directly forward, its sound waves arrive at a smaller angle, reducing the time difference between the sound hitting the two sides of the jaw. The time lag falls again if the mouse moves towards the front of the snake. And if the rodent is directly in front or behind, in true pantomime fashion, the sound reaches both jaw halves at the same time. This time difference, or lack of one, gives the snake stereo hearing and helps it work out which direction to strike.

The result? Using Rayleigh waves and simple geometry, the snake calculates exactly which direction to head for a tasty meal. Being physicists, van Hemmen and Friedel decided to model the snake's hearing, keeping things simple by pretending a Saharan sand viper's jaw is shaped like a cylinder floating in a sea of sand. They also assumed the sand is a fluid, which isn't strictly true – but what you lose in accuracy you gain in simplicity. They found the snake picks up about half the amplitude of the incoming Rayleigh sound wave; its hearing is not perfect, which is why the other half is 'lost'. Still, it's an incredible feat as that amplitude is barely a million hydrogen atoms in size. Other snakes also have stereo hearing, but none has seen such close scrutiny as the sand viper.

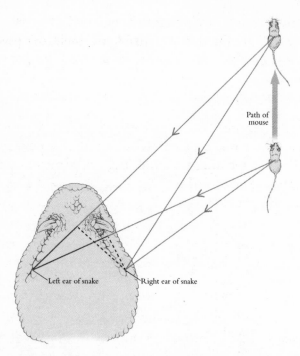

Figure 4.2 **Stereo power**. *Looking down on a Saharan sand viper, which finds prey by detecting Rayleigh waves as the animal scurries over the sand. The thick black lines show the difference in distance to the two ears, which lessens as the mouse moves off.*

We'll finish with snakes by returning to Lord Rayleigh, whose waves hold the clue to how these reptiles hear. As well as explaining why the sky is blue and bagging a Nobel prize for discovering argon gas, he wrote a two-volume masterwork *The Theory of Sound*, which he began while recuperating from rheumatic fever in Egypt in 1872–3. That country happens to be home to the Saharan sand viper, though it's not clear if Rayleigh was aware of this reptile's ability to sense vibrations. Just before he died, however, Rayleigh was president of the Society for Psychical Research, which studies the paranormal and is still active today. He would surely have been intrigued to learn that this snake's skill at detecting mice while blindfolded and unable to smell isn't psychic. Just physics.

Did the Earth move?

July 2013. Etosha National Park, Namibia. It's the dry season and the waterholes dotted across this vast expanse of savannah are teeming with life. But one African elephant (*Loxodonta africana*) calf is more alone than he'd like. Two young bull elephants charged his herd; in the panic he lost his family group. Now the anxious youngster is pacing up and down in the dust, searching for his mother. Suddenly, he freezes and lifts first a front foot and then a rear foot way up in the air.

It's a sorry scene but the calf was not totally abandoned. Caitlin O'Connell from Stanford University in the US, who has devoted much of her life to studying elephants, was watching. This was the first time she'd seen a baby elephant copy what adults do when they're troubled. Fully grown elephants shift their weight firmly onto three of their legs, raising a single foot just off the ground, in a more restrained version of the juvenile's leg-waving. Some of O'Connell's video footage approaches comedy, with the adults appearing to play musical statues, with one leg elegantly poised, as if for a photo, each time they freeze. But this wasn't a game. The elephants do their leg holds to gain more information from their surroundings. If they sense danger, the herd huddles together with the young in the middle, shielded by the bulky bodies of their elders, who have their heads up and ears held out wide. An adult male African elephant can weigh 6,000kg (6 tonnes, or 945 stone) – double the weight of a female – and be 4m (13ft) high at the shoulder, making it the largest animal that lives on land.

There was, sadly, no such heavy protection for the calf who had lost his herd. We'll return to his fate later. But why did he raise one leg when abandoned? And what's this got to do with the physics of sound?

O'Connell became beguiled by the subtle ways animals communicate when she was studying insects for her master's degree in the early 1990s. The leap from insects to elephants occurred when O'Connell headed to Africa, where she volunteered in game parks and won three years' funding through a foreign aid grant to the Namibian government to research elephants. Her background in insects was even an advantage. 'They liked that I had experience with conservation as well as pest management,' she recalls. The hope was that O'Connell could ease the problems between elephants and farmers. 'It was a no-brainer,' she says of her decision to take on the project. 'Elephants fell into my lap.' In 1992 O'Connell headed into the bush with a good-quality tape recorder, a microphone and a video camera. She's been in love with elephants ever since.

Bush telegraph

Elephants make lots of noise: they trumpet, roar, bellow and rumble. Some components of their calls are deep, with a frequency of about 20Hz, putting them in the infrasound range beyond our hearing. Elephants are loud too, but we can only put a number on it in decibels if we measure their sounds compared to something else. Named after the Scottish-born inventor of the telephone Alexander Graham Bell (1847–1922), the decibel (dB) is a tricky unit. It's a measure of the average pressure of a sound relative to a reference pressure, which we usually take to be 20 micropascals – the quietest sound a human can hear, and roughly equivalent to the whine of a mosquito flying 3m (10ft) away. So to work out the loudness of a sound, you divide its root-mean-square pressure by 20 micropascals, take the base-10 logarithm of the answer and multiply by

20. The unit is then dB SPL (sound pressure level). Like we said, it's complicated. Let's get back to elephants.

Their calls are about 120dB SPL. That doesn't quite rival the 130dB SPL volume of some bats, but it's the same as a football fan a metre away blowing a vuvuzela horn straight at you. The elephants' trumpetings travel through the air, where other elephants hear them with their large 'Dumbo' ears. If the call contains low frequencies too (around 20Hz), that sound enters the ground. This may not be down to any deliberate elephant strategy; the noise is simply so loud that its energy has to go somewhere. Like the Saharan sand viper, elephants detect these seismic components of sound waves, although, as we'll see, we're not completely sure how they do it. It's a skill shared by some reptiles, insects like the plant-hopper, and smaller mammals such as the blind mole rat, the cape mole rat and the kangaroo rat. The northern elephant seal (*Mirounga angustirostris*) is the only other large mammal known to sense seismic waves. The male seals – typically three times heavier than the females – slam their bodies onto the ground, setting up seismic waves in a 'Look how big I am, don't mess with me' display to deter other males from stealing their harem.

O'Connell reckons adult elephants form a huddle when they detect the seismic part of an alarm call rumbled out by a distant elephant that's spotted a threat such as a prowling lion. The far-off elephant is likely to be known to the group; in O'Connell's studies, elephants reacted more strongly to sounds recorded from animals they were familiar with. Seismic waves also help elephants track the movement of other elephants and even tell the mood of an approaching group through whether the individuals are walking or running. And, because other animals walk differently from the elephant's 'signature footfall', an elephant knows if, say, an antelope's wandering nearby. They may even be able to sense the approach of an earthquake – more on this later.

Hear hear

So how do they do it? Researchers aren't entirely sure. The African elephant definitely picks up seismic sound waves through the dense cushion of fat in its feet. Then it senses these vibrations either with its ears or with pressure sensors. It could even deploy both techniques – the jury's still out. If an elephant uses its ears to detect seismic sound through 'bone conduction', that might be why it freezes, leans forward and presses its front feet onto the ground. In this posture, the elephant's legs lie right below its head, transmitting the seismic vibrations directly from its feet through its leg bones to its ears, where they shake the malleus bone of the middle ear back and forth.

The other – or maybe the additional – way an elephant may hear seismic waves is through sensors in the front and back of its foot that detect tiny changes in pressure and send nerve signals to the brain. They're known as Pacinian corpuscles after nineteenth-century Italian anatomist Filippo Pacini (1812–1883), who found them in human skin, and could explain why elephants 'listening' for seismic signals lean forward onto their toes or rock back onto their heels. 'This would always happen before another family group of elephants or even a vehicle arrived,' O'Connell explains. An elephant may also feel the ground with the tip of its trunk, which contains Pacinian corpuscles too.

Either of these detection methods – ears or Pacinian corpuscles – accounts for elephants leaning forward. But why lift a foot off the ground? It seems an odd move for an elephant 'listening' out for danger. If you pick up seismic waves with your feet, surely it makes sense to plant all four squarely on the ground? But according to O'Connell, raising one foot puts more weight on the others, helping them pick up vibrations from the ground. And having three feet in place rather than four could make it easier for the elephant to work out where the sound is coming from using triangulation – a clever technique we'll return to later.

Hearing high and low

Elephants use seismic communication 'because they can,' O'Connell says. 'The signal's in the ground, it's available and they can detect it.' Seismic transmission is extra-useful when it's windy or the animals are in a forest, both of which can play havoc with sound travelling through the air. But there's another reason why seismic waves are so handy.

As we mentioned when discussing peacocks and bats, sound travelling through air spreads out in three dimensions, losing roughly 6dB each time it doubles the distance it's gone. Sound moving through the ground, in contrast, transmits mostly as a Rayleigh wave, spreading much less thinly (in two dimensions) and attenuating by just 3dB every time it doubles the distance. In other words, sound travels further through the ground before petering out than it does through air. So seismic waves give an elephant vital information from a long way off that it would miss if it only listened to airborne sound.

Like the Saharan sand viper, which also senses seismic sound, it's useful for an elephant to know where that sound is coming from. Here its dual ground/air hearing powers help. But just because sound can travel further through the ground than in air, doesn't mean it goes faster. Its speed depends on local conditions. In Etosha National Park, where the soils are sandy, it dawdles along at about 240m (790ft) per second or roughly a third less than the 343m (1,125ft) per second of sound in air (the precise value depends on how warm and moist the air is). So sound from the same source arrives at the elephant at different times depending on the route it's taken – first for the high road (air), and later for the low road (ground). Just as the time between the arrival of a lightning flash and the rumble of thunder tells us how distant a storm is, so the gap between sensing airborne and ground-borne signals lets the elephant work out how far the sound has travelled.

Elephants can also pin down where a sound is coming from more accurately by triangulation. Navigators have used this geometrical trick for centuries to estimate distances to ships, mountains or cliff tops, but elephants may have got there first. Since the elephant's two ears and the source of the sound form three points on a long, thin triangle, to work out the sound's origin all the animal has to do is sense the precise angle the sound waves make when they arrive through the air at each ear. Knowing the two angles and the distance between its ears (roughly half a metre, 20in), the elephant can figure out from which direction and how far the sound has come. Here too, seismic waves can be better than airborne sound. Since the elephant's front and back feet are about 2.5m (8ft) apart – a gap five times greater than the distance between its ears – detecting sound with its feet gives the animal a more accurate triangulation estimate of the source.

Underground sound has yet another benefit. As the elephant's feet are spaced apart, a seismic wave is likely to arrive at its front right foot at a different phase to when it reaches its front left foot. Phase describes how far a sound wave is through its cycle of high and low pressure. If the waves are out of phase, one foot will experience low pressure while the other feels higher pressure. The difference in phase is tiny, but the elephant stands a better chance of detecting it with seismic waves than with airborne sound that hits its ears. A 20Hz sound wave travelling through the air in Etosha National Park at 340m (1,115ft) per second has a wavelength of 17m (56ft) (speed equals frequency times wavelength). But in the ground, where sound travels more slowly, the wavelength is just 12m (39ft). Waves that have gone through more cycles, because they have a smaller wavelength, will be more noticeably out of synch, helping the elephant locate their source. What's more, if it's using its feet, the elephant has up to four sensors available, or even five if you count its trunk, not just two. To modify the famous line from George Orwell's *Animal Farm*, it could be 'four legs good, two ears bad'.

While these sound-localising powers could potentially have brought the lost calf that O'Connell spotted a happy ending, in this case, they didn't. But all was not lost. Elephants are social creatures, and a group of older, unrelated males waited nearby until he gave a loud bellow. Then his family returned, probably after picking up his cry through the air.

According to rumour, seismic detection also gives elephants an early warning of earthquakes. But the evidence from the devastating Indian Ocean earthquake and tsunami of 2004 is mixed. There are many tales of Asian elephants moving towards high ground before the tsunami hit, saving the tourists riding on their backs. Yet the only scientific evidence, O'Connell says, is from an elephant group in Sri Lanka that happened to be wearing satellite collars. And these animals didn't react. 'Some anecdotes match what we would expect,' adds O'Connell, with elephants responding to the ground movement about an hour before the wave struck. She believes it's possible that a barrier shelf in the ocean stopped earthquake vibrations from reaching the Sri Lankan group. 'Anecdotes are powerful if you can build statistics on them,' O'Connell adds.

Parks and recreation

Elephants' seismic communications are so good that we tap into them ourselves to find stray animals and return them to safety. Male elephants roam far and wide when seeking a female who's ready to mate. It's a time-critical business as female elephants' eggs are ripe for fertilising for only four to five days every four to six years. The males' wanderings can lead them from the safety of a national park and into dangerous confrontations with humans living nearby. Park rangers usually round up stray elephants with helicopters or by firing gunshots into the air. But by giving absent males a blast of a pre-recorded call – containing both airborne and seismic components – from a female ready to mate, staff could tempt elephantine Romeos back to a protected area. 'The sooner you can get them back, the better,' says O'Connell. To see the resemblance between speeded-up video footage of a male elephant listening out for a female and Charlie Chaplin's trademark walk with cane and oversized shoes, check out O'Connell's website at Utopia Scientific.

The beauty of using the seismic parts of the call as well as the airborne components is that you can lure males from further away – several kilometres, rather than the 500m (1,650ft) for audible broadcasts. If rangers gradually move the source of the recording back towards the park while keeping at a safe distance, they can relocate the elephant without danger to themselves. What's more, by detecting the seismic waves made as the animals walk, conservationists can count elephant numbers or spot illegal poaching.

Humans can detect vibrations through the ground too, though not well enough to capture elephants without help from machines. Native Americans may have located large herds of bison with the pressure sensors in their bare feet, while didgeridoos played by the indigenous peoples of Australia create some of their musical effect by vibrating the ground – one end of the instrument rests on the floor. Some people with limited hearing use the part of the brain that normally handles signals from our ears – the auditory cortex – to analyse vibrations instead. When O'Connell's not in Africa, she exploits this ability to make a 'vibrotactile' hearing aid that transmits useful sounds like a doorbell or phone by vibrating against the skin of your hand. 'We're trying to develop a language like Braille,' says O'Connell, who reckons the aid will also help train the auditory cortex in people struggling with their cochlear implants (devices that replace the cochlea's sensory hair cells by converting sound signals to electrical pulses). For the non-hearing-impaired, vibrations are not so key these days. 'We can perceive vibrations but we don't pay attention,' says O'Connell. 'We have cell phones for long-distance communication.'

From insects to elephants to hearing aids has been a remarkable journey for O'Connell, who finds working with these animals magical. 'I get to go out into a remote environment that's closed to the public, see individuals I know, watch them communicate and learn something new every day,' she says. 'I feel privileged to have a window into elephant society. It would take many lifetimes to figure them out.'

Sounds fishy

And so from one end of the animal scale to another. Now we don't know if you're one of them, but some people swear that if you cook a lobster in a pan of boiling water, you'll hear it scream as it dies. The noise, they'll say, is evidence that the animal's last minutes are agony. Zoologists can't be sure lobsters feel pain but any sounds it emits are from air bubbles trapped between the crustacean's shell and body that expand in the heat and squeal as they're forced out between gaps in the shell.

That's not to say lobsters are mute when they're alive. Go to the University of Rhode Island's 'Discovery of sound in the sea' website and you can play recordings of all sorts of underwater animals, from lobsters, whales and shrimp to barred grunts, bigeye scads and dusky damselfish. Many make noises that landlubbers will find weird. One – the oyster toadfish – sounds like a demented goat. But from a physics point of view, the California spiny lobster (*Panulirus interruptus*) is top of the strange-sounds charts.

Reddish-brown and up to 30cm (12in) long, this crustacean has natty vertically-striped legs and enormous, spiny antennae. It lives off the Pacific coast from California down to Mexico. The lobster makes a noise like someone rasping the teeth of a metal comb over the edge of a table in short, sharp bursts. Stranger still is how the California spiny lobster creates that sound in the first place.

According to Sheila Patek from Duke University, the US biologist we met in Chapter 2 for her work on harlequin mantis shrimp and trap-jaw ants, the ocean teems with a wild cacophony of pops, crackles, whistles and buzzes. Breaking waves, raindrops, earthquakes, ships, submarines and military sonar devices rumble, patter, swoosh and pulse. Animals, including the bubble-creating pistol shrimp from Chapter 2, make noises to communicate, attract mates, defend their territory, sense their surroundings and find food. Light, as we'll discover in Chapter 6, doesn't penetrate far below the surface of the sea. Deep down, it's murky or pitch black, and for many marine animals, sound is king.

Underwater, sound is a whole new kettle of fish. 'It's fundamentally different,' says Patek. Sound travels much further than through the air because water molecules don't lose much energy as they vibrate. With so little attenuation, noise from sea creatures goes thousands of metres before fizzling out, making it hard for the animals to hide. There's another problem too: sea-dwellers wanting to make out a message must sort through a jumble of sounds, confusing echoes and background noise.

What's more, sound travels five times faster in water than through air. To see why, let's go back 350 years to our resident genius Isaac Newton, who has made it into every chapter of *Furry Logic* so far. Brilliant though he was, Newton got it badly wrong when calculating the speed of sound. His fellow seventeenth-century scientists had measured sound's speed in air fairly well despite lacking any high-tech kit such as accurate watches. Some tested how far they had to stand from a wall to hear an echo in, say, half a second, while others watched for the flash of a distant shotgun before using a pendulum to time how much later the noise reached them.

By around 1660 these pioneers agreed that sound moves through air at 349m (1,145ft) per second – a shade over today's accepted figure of 343m (1,125ft) per second in dry air at 20°C. So far so good. But in his 1687 blockbuster *Principia Mathematica*, which contained details of his laws of motion (see Chapter 2), Newton included an equation that, he claimed, predicted the speed of sound in any medium. The problem was, it gave a speed in air that was 20 per cent *lower* than the value measured by listening for echoes or guns. Gadzooks! Surely the brilliant Newton hadn't made a mistake?

Our genius thought that to calculate the speed of sound you divide the pressure acting on the medium that the sound is travelling through by the medium's density, then take the square root of the answer. The fool! Although Newton fiddled his calculations to try to match the measurements, his theory was not, ahem, as sound as he'd hoped. The mismatch remained until the early nineteenth century when French mathematician Pierre-Simon Laplace (1749–1827) spotted a subtle but vital error in Newton's thinking.

Newton knew a sound wave consists of alternating regions of low and high pressure, but he hadn't twigged that compressed regions warm up slightly as the molecules rub against each other. The sound vibrations happen so fast that the heat can't escape. The higher temperature leads to a higher pressure and so a faster speed for sound. Take all that into account and out pops an equation now known as the 'Newton–Laplace' formula. This says that to calculate the speed of sound in a medium you divide its stiffness (not the pressure acting on it, as Newton thought) by its density then take the square root of the answer. Plug in the numbers and that bothersome 20 per cent discrepancy disappears. For this and other efforts, Laplace is one of the 72 French scientists, engineers and mathematicians to have their names engraved on the Eiffel Tower (you'll see his on the north-west side just below the first balcony, if you're not too busy eating chocolate crêpes from the stall nearby).

Fast-forward to the twenty-first century and we can use this double-barrelled formula to work out the speed of sound in the ocean. Seawater has a bulk modulus – a measure of stiffness – of 2.2 billion N/m^2 (newtons per square metre) and a density of 1,025kg/m^3 (kilograms per cubic metre) at 20°C. Divide the first number by the second, take the square root, and you get ... er, taps numbers into calculator ... a speed of 1,500m (4,900ft) per second, rounded up to the nearest hundred. That's five times sound's speed in air, give or take a little depending on how warm, salty or deep the water is.

That high speed is more than a curiosity; it makes a huge difference to life underwater. On the downside, sound's swifter passage means animals must react faster to know where a sound's coming from. But the upside is that it's easier to detect sound right up close to a source. In this 'near-field' region, the water molecules carrying the sound vibrate much more strongly than molecules further away. These powerful, up-close vibrations are also present for sound in air, but because sound travels five times faster underwater, in the sea they extend five times further than above ground.

Lobsters, along with crabs, shrimps and other marine crustaceans, use hairs on their legs to 'feel' these near-field vibrations. They can sense sound without needing ears like ours, which have an eardrum that picks up the rise and fall in pressure of a sound wave, rather than feeling the vibrations of the molecules directly. 'If I were to pass you on a sidewalk maybe I would feel a brush of air from you,' says Patek. 'But underwater you're sending me a complete set of frequencies and wavelengths that go along with your motion.'

Softly, softly

Not only does the California spiny lobster hear differently from us, it also generates sounds in a way that Patek thinks is unique. It has to. Every couple of months, this crustacean throws off its hard exoskeleton in a moult that's vital if the animal is to grow. But losing its armoury comes at a price. A naked squidgy body is a juicy snack for a passing fish, octopus or sea otter, and the California spiny lobster doesn't have claws to defend itself, just spiny antennae. So this lobster uses sound against its enemies. Unlike many other arthropods, however, the animal can't make a buzzing noise by rubbing something hard over a series of bumps, like strumming a thimble on a washboard. This 'pick-and-file' method needs two hard surfaces, exactly what the spiny lobster doesn't have when it's shed its shell.

Instead, as Patek has found, the spiny lobster is the first animal we know of that creates sound by using its body as a violin. Yet the noise is far from melodious: it's made to scare. Less Beethoven concerto, more car alarm. It uses a technique known as 'stick and slip', which involves rubbing a soft surface against something smooth. For the violin that's a bow, traditionally made from the tail-hair of a grey horse, rubbing against a string (nylon if you're modern or catgut for traditionalists – nothing to do with cats, but the insides of a sheep or goat).

'With a violin you have constant bowing with the arm going up and down, but if you took a much closer look with a high-speed video, you'd see that the bow is periodically sticking and slipping relative to the string to generate that vibration,' says Patek. That's why violinists smear rosin on their bow – it helps it stick. As the bow 'glides' across the string, it sticks then unsticks and slips, over and over again, generating a vibration – and hence a sound – each time it moves.

The lobster doesn't have a bow, but it does have a soft plectrum on the end of each antenna that it rubs over a smooth, stiff 'plate' below each eye featuring what Patek calls 'frictional shingles, like the roof of a house'. Every time the plectrum jerks over the plate it makes a noise. The lobster uses both antennae to make sound at the same time, like a violin player with two bows. When the crustacean moults, the stiff plates beneath its eyes soften but its sound system still works because it doesn't rely on hard rubbing parts, unlike the washboard technique used by insects such as crickets. 'This stick-and-slip system with the two soft surfaces is great all the way through the moulting cycle and the spiny lobsters use it avidly when they're at their most vulnerable,' says Patek.

The system relies on friction, which we met in Chapter 2 – it's the sticking force when two surfaces move against each other (the one that confused ancient Greek philosophers by diverting some of the energy of a rolling cartwheel). In stick and slip, friction takes energy from the motion of the lobster's antennae or the violinist's arm and converts it into sound. For a concert violinist, the movement becomes music. For an unskilled child, it's transformed into a painful caterwaul, and for the lobster into an off-putting shriek.

If played properly, the violin sounds good because its strings and body are designed to resonate at certain frequencies. The California spiny lobster, on the other hand, generates a wide range of frequencies. 'It doesn't sound flutey,' says Patek, 'it's just a raspy broad noise, which is what most anti-predator sounds are like. They're not trying to hit a sweet spot on the ear, they just want to blast predators' hearing systems.' Like any animal on land that makes an obnoxious noise when under stress, such as birds that make a 'pishing' noise when mobbing a cat or a captured rabbit that screams, the lobster shrieks as loudly as it can in the hope that its attacker will find the noise painful and leave.

You can hear the California spiny lobster at Patek's website; the rasping sound made by stick and slip is surprisingly harsh given that one of the surfaces creating it is soft. The other lobsters at the site, which make noise in a similar way, are worth a listen too. *Palinustus waguensis*, which hangs out in the western and central Indian Ocean, sounds like corn popping in a metal saucepan. Its neighbour *Panulirus longipes* appears to have taken inspiration from the kazoo. Other lobsters could be mistaken for frogs.

Like many things in life, using noise against your enemies isn't straightforward. 'There's a catch-22,' says Patek. 'You want to make a sound to scare away a predator and make them drop you and run away, but at the same time you've also let all the animals around you know that you're about to be attacked and they may come in and eat you.' As a lobster defends itself from one aggressor, it highlights its vulnerability to others. However, the lobster has lucked out. Underwater you'd expect the lobster's 'shrieks' to travel a long way, but Patek's measurements show that the ambient noise is so loud that these anti-predator warnings only travel about a metre (3ft) before they're lost in the background.

There's still much we don't know about how stick-and-slip friction works. 'It's a really interesting physical mechanism and a hugely difficult area of research,' says Patek. 'There's a laundry list – little is known about sounds produced by any animals in the ocean outside of marine mammals and some fish, so placing this in a broader context is challenging; the sounds definitely [move] through the water; however, they may also propagate through the seabed; we don't even know whether most marine animals have ears or not … or whether they exclusively hear in the near-field; and finally, the ocean is a noisy, variable place and making clean, realistic measurements is almost impossible.'

We might not fully understand the physics of how the California spiny lobster makes a noise, but one thing is clear: the sound puts predators off. In one study, spiny lobsters that researchers had prevented from using stick and slip were attacked faster than those that could still 'shriek'. By playing the violin, the lobster claws back self-defence.

A sound summary

It's remarkable that sound – an ethereal wave of high and low pressure – can have such profound effects throughout the animal world. For humans, sound is often an irritant: the noisy students having too much fun, the cars thundering past the house, the washing machine spinning like a banshee. We'd all cope just fine without those noises. But for many animals, sound is crucial for survival. As we've seen, peacocks use infrasound to woo partners. At the other end of the spectrum, bats seek prey with ultrasound and use it to avoid bumping into cave walls. The Saharan sand viper tracks mice from vibrations moving through the ground, while elephants sense danger that way too. As for underwater animals like the California spiny lobster, sound is vital for self-defence.

Our sonic journey has also taken us through the physics of sound: speed, wavelength, frequency, amplitude and pitch. We even looked at attenuation, triangulation and sound underwater. In Chapter 6, we'll come to another kind of wave that's crucial to animals: light. But first let's explore two physics phenomena that mystified humans for centuries. We'll start with the shocking South American fish that triggered our understanding of both.

CHAPTER FIVE

Electricity and Magnetism: Let the Sparks Fly

TASER EELS * THE CASE OF THE CHARGED BEES * TURTLES
THAT LOOP THE ATLANTIC * HORNETS SKILLED IN
QUANTUM MECHANICS

Living in electric dreams

When Tim Berners-Lee dreamt up the World Wide Web while working at the CERN particle-physics lab near Geneva in the early 1990s, the idea was to help physicists across the globe share scientific research data. He could never have predicted it would one day lead to people streaming endless cute animal videos on YouTube. One online sensation has been Tardar Sauce – 'the world's grumpiest cat' – who went viral in 2012 after her owner posted footage of this miserable-looking sourpuss. You can even buy mugs, T-shirts, books and calendars featuring the discontented feline.

Over on Twitter, you'll find enthusiasts tweeting about everything from ducks and dogs to parrots and guinea pigs. Our favourite tweeting animal, though, is Miguel Wattson, an electric eel (*Electrophorus electricus*), who lives in the Rivers of the World gallery at the Tennessee Aquarium in Chattanooga, US. He sent his first tweet in late 2014 under the handle @EelectricMiguel and although aquarium staff write his messages, Miguel controls when the tweets go out. Gadgets pick up the electrical pulses Miguel emits while roaming around his tank and a new tweet from a pre-written list hits social media if a pulse is above a certain strength.

The tweets are mostly groan-inducing animal jokes, such as 'What is the strongest creature in the sea? A mussel!', 'What do you call it when it rains chickens and ducks? Fowl weather!' and 'Why do hummingbirds hum? Because they can't remember the words!' Meanwhile, an amplifier converts Miguel's pulses into a stream of 'pops' that visitors hear as they wander through the aquarium. Each pop also triggers a light bulb, the brightness of which depends on the strength of the pulse – turning Miguel's otherwise silent electrical emissions into a *son-et-lumière* feast.

Terrible puns aside, the project does have an educational side: to raise the profile of electric eels. These animals don't have the best reputation. Their electric pulses, which they use to stun and kill crabs and other prey, are strong enough to injure humans. So if you're planning to visit Miguel Wattson, do keep your hands out of his tank. That said, as long as you keep your distance, electric eels are the perfect candidate to kick off this chapter on animals' use of electricity and magnetism. Back in the late eighteenth century, these creatures inspired some of the earliest research into this branch of physics. What's more, researchers recently discovered that electric eels stun their prey in a similar way to the Taser weapons used by police. These 'guns' were invented by NASA physicist Jack Cover, who coined the name in honour of a character from a children's book: it's an acronym of Thomas A. Swift's Electric Rifle.

We should point out that, despite appearances, electric eels aren't eels at all. They're bony, freshwater fish that mooch about at the bottom of murky swamps and creeks along the Amazon and Orinoco rivers in South America (eels are fish too but from a different grouping). Grey-brown on the back and yellow-orange on the belly, electric eels are up to 2.5m (8ft) long, with a flat head and wide mouth. Their eyesight is terrible, and they're mostly out and about at night, swimming by wobbling the single, long fin that runs down their underside. The river waters contain so little oxygen that electric eels can't get enough through their gills; they rise to the surface every 10 minutes to breathe air before sinking back to the river bed. They have strange reproductive habits too: the female lays her eggs in a nest the male makes from saliva.

R-eel-y scary

But it's their ability to create electrical pulses for which electric eels are most famous – and feared. News of these mysterious creatures first reached Europe during the sixteenth and seventeenth centuries. French astronomer Jean Richer (1630–96), for example, reported seeing an eel-like fish on an expedition in 1670 that was 'as fat as a leg' and had numbed his arm for 15 minutes when he touched it with his finger. Intrigued by these 'trembling eels', European travellers carried out some unscrupulous experiments that today would be banned on ethical and health-and-safety grounds. They'd 'invite' local people to stick their arms into a tank containing one of these fish and touch it with their bare hands or a metal stick. It was incredibly dangerous – we now know that a fully grown adult eel can create an electric shock of over 600 volts (V) – and few wanted to repeat the test. When an English surgeon called Dale Ingram tried to pat an electric eel with an iron hoop from a Madeira barrel in 1745, the fish's strike was so strong it knocked the entire metal object from his hand, like an invisible opponent disarming a fencer.

As historian William Turkel describes in his book *Spark from the Deep*, such accounts caught the ear of eighteenth-century scientists such as America's Benjamin Franklin and Britain's Joseph Priestley, who were seeking to uncover the mysterious phenomenon of electricity. The stories persuaded these investigators that the trembling eel's shock was electric in origin – a suspicion confirmed in 1775 by British scientist John Walsh, who found that an eel, imported to London from South America, could create a spark of electricity through the air. The sighting convinced doubters that the shock from the animal was linked to more familiar forms of electricity such as lightning.

With our laptops and light bulbs, these days we take electricity for granted. But back in the eighteenth century it was a strange phenomenon, with public displays of electrical effects being hugely popular. German professor Johann Heinrich Winckler, for example, electrified a servant before offering him a glass of brandy. 'Sparks from the servant's tongue ignited the brandy to the amusement of everyone present, probably excepting the servant,' writes Turkel in *Spark from the Deep*. Such demonstrations had a more sinister side too. Swiss medic Jean Jallabert used a Leyden jar – a device for storing electric charge – to stimulate involuntary contractions in the muscles of his own arm, triggering a craze for 'electrotherapy'. Doctors wanted to find out if electricity could cure patients of everything from epilepsy and lockjaw to rheumatism and toothache. Turkel tells the horrific story of an official in the Dutch colony of Essequibo (in present-day Guyana), who every day would throw a slave boy who had crooked arms and legs into a tub containing an electric eel. The fish shocked the boy so powerfully that he had to crawl out or be yanked free by an assistant, who received a shock too. Far from being cured, the youngster was instead left with permanently deformed shin-bones.

Those early studies were unsavoury, but electric eels had a huge scientific impact. News of John Walsh's proof that eels make electric sparks spread quickly across Europe, inspiring Italian scientist Luigi Galvani (1737–98) to use frogs to show that electricity helps nerves and muscles to work. Electric eels, it's fair to say, triggered the birth of modern neuroscience. Walsh's work also motivated another Italian researcher, Alessandro Volta (1745–1827), to press the ends of a wire made from two different metals into the nerve of a frog, which caused its muscles to contract. When Volta put one of these bimetallic wires into his own mouth, he noticed an acid sensation and suspected the wire was electrically stimulating his tongue's taste sensors. The experience led him to build a device for creating electricity by stacking discs of silver and zinc on top of each other, separated by thin bits of paper soaked in a salt solution. This 'voltaic pile', inspired by the electric eel, was the world's first battery. Volta's device had so many parallels with this fish that when he wrote up his findings in a paper for the Royal Society in 1800, he called it 'an artificial electric organ'. We now know that an electric eel has several thousand thin, disc-like cells stacked on top of each other, like a giant version of Volta's primitive battery. Known as electrocytes, these discs sit in three dedicated organs that make up two-thirds of the eel's body, while its heart, intestines and liver are crammed into a narrow region at the front. But how do these cells generate electricity? And while we're at it, what is electricity anyway?

Current thinking

Electricity takes its name from the Greek for amber: *electrum*. If you rub amber with animal fur, this yellow, fossilised tree sap becomes electrically charged. Rubbing transfers electrons – negatively charged particles found in any atom – from the fur to the amber. You can now use the negatively charged amber to pick up bits of paper (the amber repels electrons in the paper, making it positive on the surface and so attracted to the negatively charged lump in your hand). The same happens if you rub a balloon against your jumper. By shifting electrons to the balloon, you can amuse yourself by sticking it to the wall or making your hair stand on end (perfect for livening up children's parties). This triboelectric effect (from *tribo*, the Greek for 'I rub') can also charge you up by several thousand volts if you walk across a nylon carpet. Reach for a metal door handle and the build-up of electrons on your hand may spark through the air to the metal. Ouch!

Fur-rubbed amber or your nylon-charged body are examples of static electricity; their surfaces have more (or fewer) electrons than they ought. The charge stays where it is – it's static – unless it can flow away as an electric current, as it does between you and the door handle. Static electricity is vital for photocopiers, but in everyday life we mostly experience electricity as electrons flowing in a current down metal wires and cables or through transistors (powering your screen if you're reading *Furry Logic* as an e-book). That's what electricity is – a flow of electric charge. If the charge is in the form of electrons, we're talking big numbers. An electric current of 1 ampere (A) has more than 6 billion billion electrons passing any point in 1 second. The unit is named after French physicist André-Marie Ampère (1775–1836), who showed that two current-carrying parallel wires attract each other if the currents flow in the same direction and push apart when the currents flow in opposite directions. Like Pierre-Simon Laplace (see Chapter 4), Monsieur Ampère has his name engraved on the Eiffel Tower. As does another French physicist with a double-barrelled first name: Charles-Augustin de Coulomb (1736–1806), whose surname is given to the amount of charge transported by a current of 1 ampere in 1 second. It's crowded up there.

The French don't have a monopoly on electrical units. Italy's Signor Volta gave his name to the volt, which is a measure of electrical 'potential'. In a battery, electrons have the *potential* to flow from the negative to the positive terminal, but they'll only do so if you wire the two ends together, say via your bike light. Current is like water flowing in a mountain stream: it will flow only if you've spent energy transporting it to the top, with the height of the mountain equivalent to the voltage. If the mountain's tall, the voltage is big and water will cascade down. If it's small, so is the voltage, and the flow is slow. Stack two mountains on top of each other (if you can imagine such a thing) and you'll

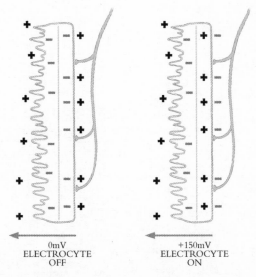

0mV
ELECTROCYTE
OFF

+150mV
ELECTROCYTE
ON

Figure 5.1 **Switched on**. *An electric eel has thousands of electrocyte cells stacked together, with nerves (curly lines) connecting one side of each cell to the brain. The voltage is zero when the cells are off, but +0.15V when on. (Based on illustrations from askanaturalist.com)*

double the voltage, like wiring up two 1.5V batteries back to back to get 3V. Multiply the current by the voltage and you get the amount of electrical power measured in watts, a unit named after Birmingham-based engineer James Watt (which is why the Tennessee electric eel is called Miguel Wattson).

An electric current doesn't have to involve flowing electrons, however. It can also consist of atoms that have lost electrons (making them positively charged ions) or gained electrons (resulting in negatively charged ions). That's what happens in an electric eel, which makes currents from positive sodium, calcium and potassium ions. These move through the surface of each electrocyte cell until they're clinging to the outside, which becomes positively charged. The inside surface of the cell – with a shortfall of positive ions – becomes negatively charged. Each side of the cell is like a tiny battery with plus and minus terminals, and a voltage of 0.085V. Not huge, given that an ordinary AA battery is rated 1.5V and you need two to power a bike light. There's another problem too: as both the left and right sides of the electrocyte are positive on the outside and negative inside, it's like having two batteries with the negative terminals touching (+ − − +). The overall voltage is zero and no current flows. But wait ... the electric eel has a trick up its sleeve.

Each electrocyte isn't symmetrical: one side is knobbly, while the other is smooth and connected by nerve fibres to the eel's brain. When the eel thinks electric thoughts, tiny pores on the smooth side open, allowing the positive ions clinging to the outside of the electrocyte to stream back into the cell. The smooth side is now positive on its inner surface and negative outside, creating a voltage of 0.065V on that side of the cell. The knobbly side is still at 0.085V, positive on the outside and negative inside, so the cell is + − + −. Add the two numbers and you get 0.15V, which is much more powerful.

Eventually the pores close and the cell loses its voltage, but the electric eel's no fool and it synchronises its electrocytes so they're all 'on' at the same time. With as many as 6,000 cells, an eel can generate a voltage of 6,000 × 0.15V = 900V. The fish is, in other words, like a giant 900V battery with a positive terminal in its head and a negative terminal in its tail. Current flows from its front end, through the water, and back to its tail. It's hard to say how big the current is because it flows through the whole space around the eel. Any animal that crosses the fish's path, however, will be in for a shock.

Zap! Pow!

That's how an electric eel generates electricity, but how does it use these currents to stun its prey? Despite countless studies of these fish over the centuries, it's a question that remained unanswered until 2014, when Kenneth Catania from Vanderbilt University in Tennessee took up an interest. Anyone wanting to experiment on an electric eel first needs to get their hands on one of these slippery customers. Catching them is hard because you must make the fish discharge their electricity until they're worn out. Instead, Catania bought four eels from a commercial supplier and housed them in a plastic aquarium the size of a supermarket trolley. To make the fish at home, he added gravel, plastic branches and plants, while ensuring the water was a cosy 25–26°C. Ernie, Ellie and his two other eels (who remained nameless) were fed earthworms and crayfish.

Keen to watch his quartet in slow motion, Catania rigged up a high-speed video camera and installed electrical sensors in the water to measure the electric currents the fish produce. Catania noticed, as others have done, that electric eels are no slouches. These fish can spot suspicious changes in the flow of water around them barely three-hundredths of a second after they've occurred. 'Even if they're asleep and you stir the water ever so slightly they'll immediately wake up,' Catania explains. Electric eels know what's around them thanks to small, pit-shaped sensors on their faces that constantly monitor the electric field in the surrounding water. If you're mad enough to dunk your hand in next to one of these eels, you'll distort the field, which changes the amount of current flowing across the eel's skin. It's a tiny effect, but the eel's sensors are super-sharp, making these fish masters of 'electrolocation' (not to be confused with echolocation – that was bats in Chapter 4). When they sense a tasty victim, electric eels don't hang about – they can strike and swallow their dinner in a tenth of a second.

Electric eels don't just wait for changes in the electric field. Except when they're having a nap, these fish are always on active duty, navigating through murky river waters, sending out low-voltage pulses themselves. When an eel suspects there's something tasty moving through the water or lurking in nearby vegetation, it fires bursts of two or three pulses at a much higher voltage. If that something is an animal, its muscles will twitch, inadvertently creating tiny ripples in the water. Thanks to those sensors on its face, the eel knows if the object is edible; a bit of wood, for

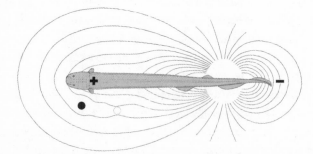

Figure 5.2 **Detecting trick**. *An electrically conducting object (black dot) moves the electric-field lines right at the surface of an electric eel closer together, boosting the electric current across its skin, while an insulating object (white dot) spaces the lines apart, lowering the current. (Adapted from Henning Scheich)*

example, wouldn't move a muscle and wouldn't disturb the water in response to the eel's test shock.

Roughly 20–40 milliseconds after sussing out that its target is alive, and with dinner firmly in its sights, the electric eel now blasts out a machine-gun-like volley of high-voltage electrical pulses at 400 times a second. At up to 500V, these pulses aren't much fun for the eel's target; within milliseconds of being hit by the first one, the animal can't move. The pulses act on the prey's motor-neurone nerves, making the animal involuntarily contract its muscles. With its prey unable to escape, the eel can start snacking; it's as if it has a remote control over its victim's muscles. The same happens if a police officer fires a Taser at you. The weapon embeds two metal barbs into your body and gives off almost 20 pulses of high voltage per second, which will have you lying flat on your face before you know what's hit you. Unlike the eel's victims, you won't be eaten but will be carted off in a police van for questioning. Just like the effects of a Taser, the paralysis in the prey is only temporary. If it doesn't immediately get eaten, the target animal regains its power of movement and can head off to safety. Sending a couple of test pulses before the main volley is, however, 'a really clever way of saving energy', according to Catania. 'Firing off round after round of high-voltage pulses is energy intensive and the eel will only want to do this if it's sure the prey is alive,' he says.

The eel has a couple of other tricks too. Catania discovered that the eel not only uses low-voltage pulses for electrolocation, which biologists had known since the 1950s, but it deploys its high-voltage volleys for the same purpose too. In other words, the eel uses its high-voltage blasts for two purposes: to stun prey, as we saw above, and to zero in on a paralysed victim that's not stationary but floating through the water. What's more, when an electric eel attacks a large animal, such as a crayfish, it has yet another tactic. The eel bites its prey before curling its tail round the victim that it almost touches its own head.

By sandwiching its victim between its positive and negative poles (its head and tail), the eel more than doubles the strength of the electric field from its high-voltage pulses. The increased power coupled with the high pulse rate drive its prey's muscles so hard and fast that they become exhausted. With the prey worn out, the eel can swallow its entire victim in peace.

We have to admit to developing a peculiar fascination with electric eels while writing *Furry Logic*, and Catania has too. He became interested in these fish after winning a Guggenheim Foundation fellowship to write his own book about the sensory systems in a range of animals. 'I originally thought that electric eels would be a kind of break from my main work but I soon discovered that they are the most interesting animal ever – despite having been studied for centuries,' he says. 'I got totally sidetracked.' Even the naturalist Charles Darwin found electric eels a mystery, placing their electric organs under a heading 'special difficulties of the theory of natural selection' in his masterwork *On the Origin of Species*. The great man found it 'impossible to conceive by what steps these wondrous organs have been produced'. Darwin would surely have been interested to learn that electric eels aren't the unsophisticated, one-trick ponies that biologists once thought, simply bludgeoning their prey to death with high voltages. Instead, these fish are masters of manipulating their electric fields.

Quite what he'd have made of Miguel Wattson and his tweets, though, is something we'll never know.

Let the field be with you

The beauty of electric eels is that they eased us into some key electrical concepts, such as charge, voltage and current. But when discussing how these fish navigate, we slyly slipped in one term without explanation: the electric field.

The field is a concept we touched on in Chapter 3 when looking at the Navier–Stokes equation, that chunky mathematical formula that says how fast and in which direction every point in a fluid moves. That's what a field is: a quantity you can quantify at every point in space. Solve the Navier–Stokes equation with a computer and you'd get a map with thousands of arrows showing the direction the fluid's flowing in at each point, with the length of the arrow indicating its speed. Join all the arrows up and you get imaginary lines that are a handy way of visualising the fluid field, like the lines representing wind on a TV weather map. In the same way, the electric field shows the strength and direction of the electric force between objects, such as the positively charged head and the negative tail of an electric eel, and you can visualise that field with a stream of lines flowing from one end of the animal to the other.

In one way, the electric force is similar to gravity (see Chapter 2). Just as the gravitational force between two objects weakens fourfold every time you double their separation, so the electric force between two charges follows the same inverse-square rule. But there are two huge differences. First, the electric force is about a billion, billion, billion, billion times stronger than gravity. You only feel gravity's effects for massive objects, like when your mobile phone slips out of your fingers and, attracted by the gravitational pull of planet Earth, hits the tarmac and smashes its screen. Grrr. The other big difference is that gravity always pulls objects together, whereas for the electric force, it's a case of opposites attract, but likes dislike. Electrons push electrons away. Protons – positively charged particles that skulk in the nucleus of an atom – repel other protons, but woo electrons. Now if you thought electric fields were strange, let's enter another field that's odder still. It's time to talk about an animal we met in Chapter 3 that has yet more secrets to behold.

Blooming marvellous

A buff-tailed bumblebee (*Bombus terrestris*) buzzes from purple flower to purple flower in a sea of green. Some of them she ignores, moving swiftly on. At others she lands and sucks up liquid with her long tongue. The bee appears content, but there's something odd about these flowers. They all look exactly the same. No shades of colour, no natural variations. Each is perfectly round and perfectly flat. These are E-flowers, artificial petunias. Each is made from a disc of purple plastic backed by steel and stuck flat on a green wooden table sitting on a surface covered in green gaffer tape. It's like a giant's Lego set or a scene from *Teletubbies*. On top of each E-flower lies a small white plastic trough filled with liquid, while underneath each table a black cord snakes along the ground.

Then a researcher flips a switch. Nothing obvious changes except that the bee now only pauses briefly at some of the flowers she chooses, occasionally shaking her head in disgust. She's after sucrose solution acting as fake nectar, but is instead getting a mouthful of quinine compound, which tastes bitter to bees (and to us – quinine gives tonic water its flavour). Before the switch, the bee and her fellow test subjects found sugar 8 times out of 10; now they win a sweet prize only half the time.

So how did our switch-flipper, Daniel Robert, an animal-senses expert from the University of Bristol, UK, stop the bees finding what they want? And why set up this freakish artificial garden? Is Robert some kind of bee mind-control wizard, secretly conducting experiments inspired by reading *The Men Who Stare at Goats* (2004)? More fundamentally, why do real flowers provide a meal for bees in the first place?

The flowers aren't after a social call; there's no such thing as a free lunch. True, the flowers hand out food to bees in the form of protein-rich pollen (the flower's male 'seed') to nourish bee larvae, and nectar that the bees regurgitate as honey once back at the hive. It's a fact marketers don't highlight on the packaging – honey is basically bee vomit. Vomit that tastes nice, but still. Getting its pollen eaten by a bee larva, however, won't help the plant reproduce. In return for their meal, the bees must distribute some of the pollen to other flowers of the same species, rather than taking it all to the hive. Unlike plants, they're mobile and can visit many flowers in one day.

To boost the chances of a bee picking up and dropping off pollen, flowers position their male sex organs, their female sex organs and their nectar supplies all in one place, slap bang in the middle of the bloom. Depending on the species of plant, the flowers may have both male and female parts, or only one type. When a bee pops in for a sugary drink at a flower with male parts, she brushes against the flower's pollen-covered anthers. You can see these coloured blobs at the end of filaments at the centre of many flowers – even with the naked eye. As the bee feeds, thousands of pollen grains transfer themselves to her hairy body. The bee then grooms herself, transferring the grains to the pollen baskets on her legs, creating powdery yellow pantaloons. But she can't reach the pollen on parts of her back. When the bee visits a flower of the same species that has female parts, some of this pollen drops onto them, meets up with an ovule and forms a seed. It's sex, Jim, but not as we know it.

Busy bees

Basically the bee wants food and the flower wants sex, the unison of pollen and ovule. For the bee, this is true even if the flower's artificial, made by a researcher in a lab. It's time to find out what Robert was up to with his pretend petunias. 'We asked the bees, can you find the flowers where there is a sugar reward?' Robert says. And they could, with the help of electric fields.

At the start of the test, E-flowers containing sucrose were at a voltage of 30V and E-flowers with quinine solution were at 0V. By the end of 50 training visits to the electrified flowers, a bee had learned her lesson: 30V good, 0V bad. That explains her 80 per cent hit rate at finding the sugary E-flowers. To check the bees weren't memorising the locations of the E-flowers, Robert and his team had moved the flowers around randomly between bee visits. They had also cleaned them with ethanol to remove any pheromones the bees might have left behind as a signal, and with water to eliminate traces of quinine. But what if the bees were using something other than voltage to find the sweet E-flowers – perhaps smell or another subtle difference the researchers' feeble human senses couldn't detect? To catch the bees out, the team turned off the power. 'Lo and behold, when you take the voltage away they can't do it, they go back to random chance,' Robert says. 'The students were really happy at that time, they were jumping up and down in the lab.'

The bees had shown Robert and his colleagues that they use electric fields to learn and make decisions. Impressive when you consider the size of their brains. Bees can even detect the pattern of an electric field, not just its presence. In further tests, Robert created a sweet spot from an E-flower charged with an outer ring at +20V and a centre at −10V. The quinine, meanwhile, was on E-flowers – held at a consistent +20V. Again, the bees learned their lesson and were 70 per cent accurate in their search for sugar by their last 10 visits. And they couldn't repeat the trick if the voltages were turned off.

While animals with electrical powers – the ability to sense electric fields, produce electric fields or both – are relatively common underwater, the bee is the first we know of that detects electric fields in air. These insects, Robert proved, use electric-field information to decide whether or not to visit an E-flower. The researcher, who published his findings in *Science* in 2013, is thrilled to have discovered 'a new sense on an organism that has attracted attention for the good part of 2,000 years – a lot of the food we have on our table comes straight from pollination.'

Like Robert, we've also become fascinated by bees. Along with their close relatives, wasps and hornets, these stingers are our most featured animal, followed in joint second place by dogs, snakes, ants and mosquitoes. There's a joke in there somewhere, if only we could ferret it out. But this is a science book so we can't just admire things, we need to know how they work. There are two key questions. How does a flower in the wild that's not wired up to a power source give off an electric field? And how does the bee detect it in the first place?

Buzzers ready

It's not surprising that animals create electric fields: we use electrical impulses in our nerve cells and to fire our muscles. But plants don't move a muscle and don't have brains so it's not obvious how or why they they make electric fields. And, largely, they don't, although small electric currents and potentials build up when dissolved charged ions move across the plant.

The secret lies in the air around them. On a clear day, flowers are surrounded by the electric field made by the 2,000–3,000 thunderstorms that occur around the world each day. Ice crystals rubbing together at the base of storm clouds build up a negative charge so strong that it discharges to the ground. With electrons removed, there's now a net positive charge in the atmosphere. The electric potential – the height of the mountain if we refer back to our stream analogy in eels – is pretty much zero at the Earth's surface and peaks at about 300,000V roughly 30–50km (20–30 miles) above the Earth with the electric field at flower height about 100V/m. The *Petunia integrifolia* Robert modelled are about 30cm (12in) high, which means that on a sunny day the flowers are surrounded by an atmospheric electric potential of about 30V – the same voltage Robert applied to his E-flowers. The petunia's stem and roots earth the plant, however, bringing its electric potential to the same level as the soil, which in fair weather is always more negative than the atmosphere because there are so many electrons in the ground. So there's around a 30V potential difference between the flower and the air around it. The positive charge in the atmosphere also induces a negative charge on the surface of the flower, by inviting the Earth's electrons to travel through the flower and accumulate there. When it rains all bets are off; the local atmospheric electric field is around zero because water droplets conduct electricity better than air and remove the excess charge.

A hairy business

That's how a flower has an electric field. Time for question two: how does the bee detect it? 'A lot of people have noticed that pollinators are usually pretty hairy,' Robert says. 'They're nice and fluffy, which helps them be liked by people.' As well as boosting bees' popularity, the hairs could, Robert believes, help them detect electric fields. 'Before flat-screen televisions, which are nicely non-electrostatic, I remember being zapped passing the television set,' he says. 'That's the same principle. As you approach a screen or sometimes acrylic paint at an [art] exhibition, you feel the hairs on your forearm raise. They're not designed for that but you can feel the hair rising.'

Just because something is moving, however, doesn't mean it's sensing anything. 'I could put a pebble on that table, make a noise with a loudspeaker and show you with a laser that the pebble is vibrating,' Robert says. 'Does it mean that the pebble hears? Probably not. We should not fall into this trap.' Even if her hairs respond to electric fields, the bee might not use that information, which is why Robert did his E-flower experiments. 'Let the bee tell us whether she has a sense that there is an electrostatic interaction between herself and the flowers,' he says. And she did.

What exactly happens when a bee flies near an electrically charged flower is complex because the bee has a charge of her own. We've known since the 1980s that most bees are positively charged, but Robert found that about 6 per cent of bumblebees are contrary types and are negatively charged. 'Bees are like people,' as he puts it. 'Some are more positive than others.' For these positive bees, friction with air molecules as they fly must knock electrons from their surface, leaving them with a surplus of unmatched positive charge. Bees might end up negatively charged, on the other hand, by flying through air containing dust that wipes electrons off on them. It's the triboelectric effect again.

Robert's student Dom Clarke has a great claim to fame: he was the first person to measure the exact amount of charge on a bee. It's a difficult task. Given its sting, you don't want to mess about with a highly – or even slightly – charged bumble. That said, Robert has never been stung in his four years studying the insects. 'They either like me very much or they don't like me,' he says. 'I don't know which.' Bumblebees are placid – 'they're lovely guys'. But the bees still move around. Luckily a technique developed back in the nineteenth century by bookbinder-turned-scientist Michael Faraday (1791–1867) came to the rescue.

With the aid of a pewter ice pail, Faraday showed that if you hold a charged object on a stick inside a metal bucket, without it touching the sides, a charge of almost the exact same size and sign – positive or negative – appears on the bucket's outside surface. Seemingly miraculous, but it's just physics. As Arthur C. Clarke put it: 'Any sufficiently advanced technology is indistinguishable from magic.'

Since Faraday's nineteenth-century tech is no longer cutting-edge, we *can* explain it. If there's a positively charged bee inside the bucket, the little buzzer induces a negative charge of the same size on the inside of the 'Faraday pail' by attracting many of the pail's electrons towards its inside surface. As a result, the outside of the bucket becomes positively charged, again with the same amount of charge as the bee and this time with exactly the same sign. Nothing supernatural, but a neat trick all the same.

So if you can entice a bee into a bucket using some sugar, all your charge-measuring problems 'pail' into insignificance – you just need to wire up a sensitive ammeter to the bucket, not the bee. Earlier researchers had measured the charge on bees with electrometers, but using a Faraday pail makes sure you detect it all. Using the pail-ammeter combo, Clarke and Robert found that an average bumblebee carries 32 thousand-billionths of a coulomb of positive charge, equivalent to having lost 1,000–2,000 electrons. That might sound a lot, but an average-sized bee probably contains around 10 billion billion times more electrons than this – so losing a few thousand isn't a big deal. To put this in context, 1 coulomb of charge flows through an energy-efficient 9W light bulb in 25 seconds. A typical lightning bolt, meanwhile, transfers 15 coulombs to Earth.

Who's in charge?

All of this physics means that the flowers are a little negatively charged and the bees are positively charged. So when you put them together do they cancel out, leaving a small bee-shaped hole in the fabric of space-time? As so often with headlines featuring a question mark, the answer is no. There's no disappearance, but the results can be spectacular. As a bee approaches a flower, you might see a puff of yellow pollen jump onto her hindmost legs. It's the 'bee's knees' of pollen-transfer methods, like finding that an item on your shopping list leaps into your supermarket trolley under its own steam. Again, it's not magic, it's electrostatics. As the bee and pollen become closer together, the attractive force between their opposite charges grows bigger until it's large enough to overcome the pollen's weight and any adhesion to the anthers, and accelerate the flower 'seed' across the gap.

To find out in detail what happens as a bee and a flower get up close, Robert wired up the stem of a *Petunia integrifolia* to an electrode and placed the flower inside a wooden flight arena. In a video Robert shot of his experiments, the flower looks abandoned, like a lone poppy on a patch of waste-ground. Even as a bee approaches, the electric potential of its stem begins to rise. You can watch an animation of the stem potential, which gradually decreases from 10 millivolts (mV), then rises as a bee bumbles near, disappears headfirst inside the bell of a flower, stays a while, then buzzes off. The flower's voltage continues to increase for a few seconds after the bee has left, peaking at about 40mV before it slowly drops back to zero. Even though a typical bee visits for just 4 seconds, the stem potential remains more positive than before for around a minute. It all happens in a leisurely fashion, unlike a simple electric circuit where a bulb lights up as soon as you flick the switch and let the electrons flow. 'We should certainly see something happening very fast, but we don't, it's very slow,' Robert says. There's more to learn.

Following a visit from a bee, the petunia stem was around 25mV more positively charged, Robert found. After another couple of bees had dropped in it was even more positive: each time a positively charged bee visits a flower some electrons move from the flower to the bee, reducing her positive charge and lessening the negative charge of the flower. Robert doesn't yet know how much charge transfers at each visit, or why only some of it does. It's hard to measure as the instant you add an electrical test kit, you mop up all the charges.

As we saw earlier, bees detect and learn from the electric fields around flowers. But what's the advantage to a flower of giving off such a field? 'There's no point in me, a petunia, to be visited by you if for the rest of the day you're going to visit poppies,' Robert explains. It's all a matter of marketing: the plants need brand loyalty, or flower constancy, from the bees. It's crucial for the flowers that bees feed mainly from one species. To keep their bee customers loyal, the flowers must avoid false advertising. If one flower has many visitors, it'll run out of nectar and need time out to restock. Should its colourful, sweet-smelling displays tell the bees too often they're chock-full of sugary goodness when they're not, the bees may look for nectar elsewhere, from a different species of flower that could prove more reliable. 'If you go to a supermarket and for the third day in a row they don't have the organic semi-skimmed in the half-pint version, you think, hmm, I'm going to go to another supermarket,' Robert says.

Disappointing bees is the last thing a flower wants to do. But flowers can't quickly change their bee-attracting colour and scent. Instead they've evolved a physics trick to let the bees know that they're taking a break and will be back in the nectar business soon.

By altering its electric field, Robert believes the flower tells the bee it's temporarily out of stock. It's like a digital 'sold out' sign that lights up as soon as the flower's nectar supplies plummet, rather than a display board that it must lovingly hand-paint. As Robert puts it, the flower can say 'I'm the right colour, I smell nice, but for a good 5 minutes, I don't have as much nectar. Come to see me again, but don't completely change your strategy of preferring my colour and smell, go and see my friends of the same species next door.'

So a bee passing within a few centimetres of a flower that's recently had visits from other bees can probably sense that the flower has an unusually high positive charge and an altered charge pattern and decide to come back later. 'Each bee leaves a trace of itself on that flower,' says Robert, adding that the change in electric charge is 'a shadow of the presence of another bee'. The flower and bee work together to leave a speedily posted electronic message (a b-mail?) for other bees. This helps hungry bees find nectar more quickly and means flowers can maintain bee brand loyalty. Win–win.

This is especially important for honeybees which will head straight back to the hive if a patch of flowers isn't up to scratch. There they'll communicate this sorry news via a waggle dance (more on this in Chapter 6) that could divert a whole group of foragers to more promising pastures elsewhere. Bumblebees, in contrast, don't dance to pass on information or come back to their nest as regularly as honeybees – they may even stay out overnight. Tsk.

Charging ahead

Plants respond to human movement, as well as bee, movement. Robert, who is a fan of communicating his work, has a demo in which he wires up an African violet to a voltmeter and waves his arm at it. It's eerie to see a flower reacting in an 'animal-like' way – the electric potential of the violet changes as Robert's hand approaches. This is almost 'carrots scream when you pick them' territory, as told to many a vegetarian by ardent, if misinformed, meat-eaters. We're not sure how the plant generates this electric response but, again, it could be due to electrostatics.

It's early days for our knowledge of how bees detect electric fields – there's a lot we don't understand and Robert's work continues. Bees could be using their powers to sense other information besides flowers' nectar state. Maybe these insects detect the atmospheric electric potential, which increases with altitude, to tell how high they are flying. And as the water droplets in clouds conduct electricity, a bee may be able to sense electrically when a cloud is overhead, or even on its way. With this in mind, Robert is currently counting how many honeybees leave the hive under different weather conditions.

Other insects may be able to sense electric fields as well – we might even find that bacteria can too, according to Robert. So what about humans? 'If I was on *Mastermind*, I would pass that question,' he says. 'People say they are electrosensitive. I don't know, I don't think I am. I ask them which component of the field they are sensitive to and they can't tell me. But they have a feeling that they are. I'm interested but I've got plenty on my plate to do.'

Compared to how much we know of other animal senses – sight, hearing, touch, taste and smell – we're 150 years or so behind on how bees, and perhaps other land-based animals, detect electric fields. 'It's exciting times because it's all open,' Robert says. 'Suddenly you discover all sorts of questions rushing faster than answers, always a good sign in research, but sometimes you think "OK, deep breath, which one do we chase now?" You can't chase every rabbit on sight.'

Magnetic attraction

Before we got all in a buzz about bees, we mentioned that the electric force between two objects weakens by a factor of four every time you double their separation. We call the rule Coulomb's law after our old friend Monsieur Coulomb (of the Eiffel Tower), who had been experimenting with charged metal balls in his lab in Paris in the 1780s. But the rule is only true if a charged object is stationary. If it's moving, the electric force also depends on how the object is whizzing about. The bit that depends on movement is called the magnetic force.

Electricity and magnetism are two sides of the same coin, which was a revolutionary discovery when Faraday made it in 1821. Working in his lab at the Royal Institution in

London, he found that by sending an electric current down a wire and then placing a magnet nearby, he could make the wire move. This, in a nutshell, is how electric motors work. Similarly, if Faraday moved a magnet near a wire, he created a current. That's what electrical generators do, like the turbines in a power station or the old-fashioned dynamo on your bike. Having a way of making electricity on demand let us light and power much of the world. It was also great news for electric eels, which were no longer a prized source of electricity, and could skulk around South American rivers in peace.

So placing a magnet near a current-carrying wire will move the wire. But where do you get magnets from? These days it's easy (you can even buy Einstein fridge magnets online), but in ancient times you had to be living in Greece. That's because beneath the fields, forests and beaches of Magnesia in the east of the country lie black rocks made from iron oxide. As the ancient Greeks first realised, these 'lodestones' can attract pieces of iron or other lodestones (and because the rocks are from Magnesia we call them magnets). What's more, if you suspend a piece of lodestone so it can swing freely, the magnet will line up north–south. That observation led to the invention of the compass needle, which Chinese navigators were using by the twelfth century – indispensable if you're on the open sea with no coastline or stars to guide you.

It wasn't until the sixteenth century, however, that British scientist William Gilbert (1544–1603) realised that compass needles align north–south because the Earth is itself a giant magnet. This then-radical notion formed the centrepiece of his epic book *De Magnete*, which kick-started the science of magnetism. So convinced was Gilbert of his ideas that he forked out £5,000 of his own cash – a vast sum at the time – to prove them experimentally, even if he had no clue why the Earth acts as a magnet.

Most of us can't detect magnetic fields (more on this later) but we can see them by placing a sheet of paper on a magnet then sprinkling iron filings on top. The magnetic flakes will spring to attention along the lines of the magnetic field. For a rectangular magnet, the result looks like a bar with an 'electric-shock' hairdo at either end – the magnet's north and south poles – with hairs of iron filings along the sides to join up with the other haircut. Conventionally, these lines are shown flowing from the magnet's north pole to its south pole.

The Earth is similar. Magnetic field lines flow from the southern hemisphere round to the northern hemisphere. What we call the Earth's north pole is actually a magnetic south pole. That's why the north pole of a bar magnet points towards it; as with electric charge, opposites attract. At the equator, the field lines are parallel to the Earth's surface as they arch between the poles. Geeks call this an inclination angle of zero. To less geeky geeks – let's call them geographers – it's the dip angle. Moving away from the equator, this angle (whatever its name) gets bigger. At the poles, the field lines hit the Earth square on, with an inclination angle of 90°. As Gilbert pointed out in *De Magnete*, if you can sense this inclination angle you get an idea of your latitude – your distance north or south of the equator.

Gilbert didn't know what creates the Earth's magnetic field, but we do. It's the iron in them there cores. Heated by the Earth's 5,700°C solid inner core, the liquid iron alloys in the outer core circulate by convection (like the bathwater in Chapter 1). As they swirl, the rotation of the Earth sets up spinning circulation currents. This rolling boil of molten iron is aligned north–south. No-one knows exactly how this 'geodynamo' works, but the moving metal generates an electric current, which sets up a magnetic field.

It's a weedy field, mind you, despite the Earth being so big. At its feeblest, over South America, it's just 25 millionths of a tesla (T) – a unit named after Serbian-American electrical engineer Nikola Tesla (1856–1943), who first got into science after noticing sparks created while stroking his boyhood pet cat Macak. Even at its strongest, at the poles, the Earth's magnetic field is barely three times that amount – 65 millionths of a tesla. That Einstein fridge magnet you bought online has a field more than a thousand times stronger (0.01T), which is why Albert sticks to your white goods instead of flying towards the nearest pole. Magnetic resonance imaging (MRI) scanners, used in hospitals to diagnose brain tumours, cancer and other diseases, contain magnets with strengths of a whopping 1.5T. Medical staff have to be super-vigilant to keep other magnetic objects out of the scanning room to avoid them being propelled towards the imager and killing patients.

Given the Earth's magnetic field is so feeble, you might not expect it to be of much use to animals. But it is, especially if you're a turtle.

Shelling out

Sometime back in the twentieth century, probably in 1915, a crew of fishermen from the Cayman Islands was catching green turtles off the coast of northern Nicaragua. As was the custom, the men branded their initials on the reptiles and loaded them into a boat heading for Key West in the US. But there was a huge storm off the Florida Keys and the boat never made it; it capsized and the turtles escaped into the sea. A few months later, the same fishermen, again off Nicaragua, were amazed to spot their initials on two turtles in their nets; the escapees had doggedly headed back from Florida – at least 1,150km (715 miles) away – to the feeding grounds from where they were snatched.

The captain of the crew regaled the tale to Archie Carr (1909–87), a pioneering sea-turtle investigator at the University of Florida. Carr detailed the story in his 1956 book *The Windward Road: Adventures of a Naturalist on Remote Caribbean Shores*, a publication that led to the formation of the Brotherhood of the Green Turtle. Today this organisation is dubbed, less romantically but more gender-neutrally, the Sea Turtle Conservancy.

'Prior to Archie Carr's work, fishermen in many parts of
the world were aware that turtles migrated long distances
but the scientific community was not,' says Ken Lohmann
of the University of North Carolina, US, who has unravelled
many of the mysteries of turtle navigation. 'It took careful
work by Carr and others, tagging turtles and recapturing
them, to establish that the turtles did migrate long distances
and frequently returned to the same areas to nest year
after year.'

But how do the turtles find their nest sites? Carr and his
fellow turtle researchers had several theories, all of them
involving physics. To find their way, the turtles could 'ask
the Universe' and use the stars or the Sun. Or they could
use polarised light patterns (see Chapter 6). Many of these
ideas had drawbacks, however. For one thing, turtles swim
underwater so they can't see the night sky clearly. And if
they could, in many areas clouds make it hard to see stars on
a regular basis. What's more, for turtles that migrate in the
southern hemisphere, there's no pole star that remains
almost fixed in the night sky. 'Even if they did have perfect
vision and could see the star patterns, there wouldn't be any
clear indication of where true north is,' Lohmann explains.
As for navigating by the Sun, that's no use when it's dark or
the turtle is deep in the ocean where light doesn't reach.
Scientists had a hunch that turtles had shelled out on a
different tactic. Yet it wasn't until the early 1990s that
Lohmann proved what it was, even though it had been
suspected for years.

What he tort-us

Lohmann began investigating turtles after studying
direction-finding in lobsters and sea slugs. His turtle
research was meant to be a short-term project, but he's still
going strong after more than 25 years. Lohmann is clearly
a man who enjoys his job. 'Turtles are a lot of fun to work
with,' he says. 'They're charismatic animals. The hatchlings

are cute, they have big eyes and it helps, of course, that they don't bite – they're essentially defenceless.' Turtles don't remain small for ever: a nesting adult female can be over 1.2m (4ft) long and weigh more than 110kg (240lb). 'They're large and kind of prehistoric-looking, it feels like working with dinosaurs,' Lohmann adds.

Lohmann made his breakthrough with the loggerhead turtle (*Caretta caretta*), which is named after its big head. Typically, these turtles have a reddish-brown shell built from hard plates, or scutes. On their faces and the topsides of their flippers, their skin's mottled brown and yellow, while underneath is pale yellow. One of the loggerhead's largest nesting beaches lies on the east coast of Florida, more than 1,000km (620 miles) south of Lohmann's home university. On the beach Lohmann studies, there might be 17,000 turtle nests in a season over a 40km (25 mile) stretch of coastline, each nest containing 100 or more eggs. But raccoons, foxes or ghost crabs gulp down many of the eggs and Lohmann is right to say the hatchlings that survive are defenceless. Barely 5cm (2in) long, they must run a gauntlet of seabirds as they scrabble down to the sea at night using their tiny flippers to push themselves over the sand. Even if they reach the ocean, the hatchlings aren't safe – numerous fish that will happily feast on baby turtle lurk below. Only one in 4,000 hatchlings makes it to adulthood.

'Just about the worst place for a small turtle to be is in clear shallow water over a reef near land – there are lots of predatory fish looking up at the turtles and lots of seabirds looking down,' says Lohmann. The young turtles have no good way to escape. They're too buoyant to dive more than a metre below the surface so they can't get away from birds. And they swim much too slowly to escape fish. To protect themselves from these terrible odds, the hatchlings plunge headlong into the waves and swim like crazy out to sea. Once away from land, they'll be far from the clutches of these birds and fish, most of which hang out within 50km (30 miles) of the shore.

For loggerheads (in the turtle sense, not the ancient insult meaning 'idiot') born on the east coast of Florida, the place to be when you're young is in the Atlantic warm-water current system that flows around the Sargasso Sea. To get there, the hatchlings head east to pick up the Gulf Stream that sweeps north along the south-eastern US coast before eventually bringing its warm water to Britain. Initially the reptiles find their route by swimming square on into the waves approaching the shore. But as soon as they're further out to sea they must find another tactic to stay on track (more on this later). By swimming into the Gulf Stream, the turtles enter a large circular current, the North Atlantic subtropical gyre, where they stay for the first 5–10 years of their lives, bulking up on invertebrates such as snails, anemones and jellyfish. There aren't many birds to attack them, fish predators are few and far between, and they can hide in the floating seaweed.

During that time, many of the growing turtles circle the Atlantic in a 15,000km (9,000 mile) once-in-a-lifetime round trip. 'They swim and drift around the Sargasso Sea, cross over to the coast of Spain and Portugal, move south along the northern coast of Africa, and then loop back to North America,' explains Lohmann. 'By the time they get to maybe two feet in length they are safe to move back into shallow water because there are few things left that can eat them.' Only a large shark is big enough to eat a juvenile or adult turtle. After this transatlantic feat, the bulky adolescent reptile heads to juvenile feeding grounds along the north American coast or as far south as Nicaragua.

But the turtles can't just go with the flow on their transatlantic loop. It's crucial that they aren't swept out of the gyre. Too far north and they'll freeze in cold Arctic waters. Too far south and they'll be drawn into the currents of the southern hemisphere, ending up who knows where. Problem is, the currents vary from year to year. 'The fastest way to cross the ocean is to stay right in the centre of the Gulf Stream but that's very difficult as a turtle can't know where the centre is in any given year,' says Lohmann. 'Probably by sheer luck some wind up close to the centre and are carried across the Atlantic very quickly and others get pushed out into the Sargasso Sea or a little north of the gyre system and have to swim back towards the main currents.' The zoologist reckons no two turtles swim exactly the same route, which explains why some take five years and others dawdle round in 10. But how do they know which way to go? It's not as if they have a compass ... or do they?

Richard the Ironheart?

To find out if turtles can detect the Earth's weak magnetic field, Lohmann built circular tanks of water 1.22m (4ft) in diameter surrounded by loops of wire. By passing an electric current through the loops, he could reproduce the magnetic fields that exist anywhere on Earth. 'All we did was take small hatchling turtles that were beginning to migrate and put them into cloth harnesses, basically little bathing suits,' he says. 'If you release them in a pool of water they'll swim continuously.' The picture accompanying Lohmann's research paper shows a youngster decked out in a lime-green tabard like a fluorescent cycling waistcoat. More recently, the team has put the hatchlings in bathing suits of a light shade known as 'Carolina Blue', one of the University of North Carolina at Chapel Hill's school colours.

When Lohmann replicated the magnetic field on the east Florida coast, the turtles swam east, the direction that would pick up the Gulf Stream if they were in the sea rather than a super-sized paddling pool. When he reversed the magnetic field around the turtles, most of them turned and swam in the opposite direction. 'That was the initial evidence demonstrating that turtles can sense magnetic fields,' he recalls.

Unlike the Sun, stars or light used in other forms of navigation, the Earth's magnetic field is there day or night, pretty much anywhere on Earth (though magnetic rocks can confuse matters). Magnetic navigation works at the bottom of deep ocean trenches, where light can't penetrate so animals can't see the sky, and at the highest heights that a bird can soar. What's more, the field's the same whatever the weather or season, though it does change as the years advance, a phenomenon that, as we'll find out later, helped Lohmann discover more about turtle navigation.

'Young turtles inherit a set of instructions that tell them what to do when they encounter certain magnetic fields in the open ocean,' says Lohmann. There's a field near northern Portugal, for example, that makes the turtles swim towards Africa. 'You can take turtles that have never been in the ocean before and in the lab expose them to this magnetic field and they respond by swimming south,' he explains. Off the African coast, other fields divert the turtles westwards, back to the continent of their birth.

Turtles, Lohmann believes, exploit the Earth's magnetic field in two ways. 'The simplest is to use it as a compass in much the same way we do with our hand-held compasses,' he says. 'The hatchling turtles certainly do this – they can tell the direction of the field and determine if they're travelling, for example, north or south.' But a compass alone isn't enough. Before a turtle can tell in which direction to go, it needs to know where it is. Fortunately, the Earth provides a map as well as a compass. Lohmann has shown that turtles detect not only the inclination angle of the magnetic field, which varies mainly with latitude, but also the strength of the field, which changes across the globe too. By detecting two magnetic features, turtles can use the Earth's field as a kind of 'magnetic map' – a biological equivalent of a global positioning system (GPS) but based on magnetism.

Just as a map of the Earth has contour lines representing various heights above sea level, so a magnetic map has lines where the magnetic-field inclination is the same, known as 'isoclinics' (nothing to do with hospitals), and lines of equal magnetic-field strength, or 'isodynamics'. In most parts of the world isoclinics and isodynamics aren't parallel, so almost every ocean region has a slightly different magnetic field – a 'magnetic signature' with a unique combination of inclination and intensity.

Hatchling turtles and adult turtles use magnetic maps for slightly different purposes. The youngsters employ the magnetic signatures of various geographical areas as a series of navigational markers, each indicating that the turtles should change course. For them, it's all about the journey. Hatchlings need to loop round the Atlantic on a road trip that gives them time to grow before they make for a feeding ground on the American coast. They don't have a specific destination in mind; all they care about is keeping inside the current system. But after reaching adulthood, the turtles are more focused and want to get back to the very beach where they were born.

Home, sweet home

After a 15,000km (9,000 mile) journey lasting years, the turtles do something that's equally incredible, this time for its precision, not the distance involved. 'Eventually, when they're fully mature and able to nest, at about age 20, they migrate back to the same area of the coastline where they started out,' says Lohmann. Adult females journey north to the beach of their birth between May and August every two or three years, depending on how much food they're getting. They scoop out a hole with their rear flippers, lay around a hundred eggs and cover them with sand, repeating the process some 15 days later. Males may make the trip to mate with females every year, but it's hard to tell as they stay in the water.

Why journey round the Atlantic Basin, then head back to your birthplace or thereabouts? One theory is that since a turtle can't tell by casting her beady eye at a beach whether its conditions are right for producing young, it's as good a bet as any to go back to where she was born.

She's living proof that the site worked successfully at least once, so why risk an unfamiliar beach that might be too steep, rocky, muddy, full of predators, or at the wrong temperature for eggs to incubate? We don't yet know how precise the turtles are when they return. 'The turtles that nest in south Florida are genetically distinct from those that nest in north Florida and they're probably considerably more accurate than that,' says Lohmann.

So how do the turtles find their birthplace from a feeding ground hundreds of kilometres away, off the coast of North or Central America? Knowing the direction to head when you sense the magnetic signature of a particular patch of ocean is fine if you're a hatchling looping around the Atlantic. You drift and swim along your migratory route but you don't need to reach anywhere in particular until you've finished your journey. And at that point you're simply looking for a feeding ground along a long stretch of coast. But an adult female heading back to the beach of her birth must get to the right place at the right time. She needs to up her navigational game.

To discover how the adults make it home, Lohmann exploited the change in the Earth's magnetic field from year to year, known as secular variation. You can think of the Earth as containing a giant bar magnet with its poles at an angle about $10°$ off from the Earth's axis of rotation. That misalignment is why the magnetic north pole – the point where the inclination of the Earth's magnetic field is $90°$ upwards – is not at the geographic north pole. The magnetic north pole wobbles about as the movement of the molten alloys in the Earth's core changes: for the last 180 years the magnetic north pole has been heading north-west. And it's changing speed: in the early 1900s, the pole moved at about 10km (6 miles) per year but by the early 2000s it was racing along four times as fast. Every now and then – the last occasion was 780,000 years ago – the field changes its mind completely and the north pole heads south. But it's the more recent variations that concern us here.

Over the last 20 years on the east coast of Florida, the isolines – lines where the field is of either a certain strength or inclination – have gradually shifted north. In this region, both the isoclinics and isodynamics lie roughly east–west, because the Earth's magnetic field becomes stronger and its inclination angle becomes steeper as you head north along the coast. But the lines haven't shifted northwards uniformly; in some places they've moved more than others. If there's a stretch of coastline where the isoline to the south has travelled a long way north, but the isoline to the north has only budged a little, then the area of beach between those isolines will have shrunk. So the magnetic signatures that fall between the isolines will be squashed into a smaller space. On the other hand, if the isoline to the north of a point has shifted a long way but the isoline to the south has hardly strayed, the area of beach between those isolines will be bigger and the magnetic signatures within them will span a larger patch of coast.

Flipping marvellous

When Lohmann and colleagues analysed turtle nest sites along the east coast of Florida from 1993 to 2011, they found that in areas where the isoclinics (those lines of equal magnetic-field inclination) had converged, turtle nests were nestled more closely together. And in regions where the isoclinics had drifted apart, the nests were further apart too. As for the isodynamics (the lines of equal magnetic-field strength), in Florida they had only moved further apart, with the nest sites becoming more distant the more the isodynamics had diverged. This proves, Lohmann believes, that turtles locate the beach of their birth by its unique magnetic signature.

Hatchling turtles imprint on the beach's magnetic field strength and inclination as they leave, just as a chick imprints on the first animal it sees as its mother. 'Young turtles learn the magnetic signature of their home beach, retain that information,

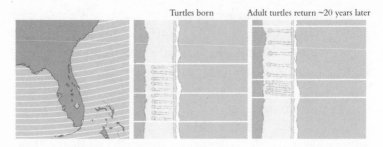

Figure 5.3 **Returning turtles**. *The diagram on the left shows the magnetic inclination isolines off Florida. In the middle are turtle nests (small circles) on a beach on Florida's east coast (with tracks trailing eastwards to the sea). On the right, it's around 20 years later and the inclination isolines have shifted north at different rates.*

and then use it as adults to navigate back years later,' Lohmann says. When they return in 20 years' time, some turtles have hit the real-estate jackpot as the area of beach they see as home has grown under the Earth's magnetic field changes and there is plenty of room for nests. Others are not so lucky. Their magnetic homeland has shrunk and they must lay their eggs close to others. Magnetic imprinting on their birthplace by turtle hatchlings. QED. Once again Lohmann had solved a mystery that had intrigued turtle researchers for years.

It's all very well knowing the magnetic signature of where you want to be, but how do you find it? Near the coast of Florida the isoclinics and isodynamics line up roughly with latitude, so it could be hard for a turtle to work out its longitude (its location east–west). But since the coast, and many of the other seashores that make excellent turtle nest habitats, such as the west coast of Africa, runs north–south, this is OK. Once the turtle has found the coast she can swim parallel to it, heading north or south as the magnetic field inclination angle and/or strength reveal she's getting magnetically 'warmer' or 'cooler' in her hunt for an exact magnetic match for her birthplace.

How turtles detect magnetic field strength and inclination is one mystery Lohmann hasn't yet solved. But he's working on it. That makes magnetic sensing, along with the bee's ability to detect electric fields, unusual; biologists more or less know how animals see, hear, smell, taste and touch. One idea is that turtles have small crystals of magnetite, the iron oxide found in compass needles, in their brains or linked to their nervous systems. As the magnetite aligns with the Earth's magnetic field, it could press on secondary receptors, signalling to the brain which way's north. This makes sense, but there's a catch – while studies in the 1990s found magnetic material inside sea turtles, there's not enough proof. Lohmann is hunting for crystals of magnetite in turtles using MRI scanners. He's not found any so far. 'If they exist they're very small,' he says. 'If you have a tiny little particle creating a tiny little magnetic field that may not be enough to be detectable with MRI.'

The other idea of how turtles navigate by sensing magnetic fields is that they use pigment molecules called cryptochromes. Sensitive to blue light, they're found in the retina at the back of the eyeballs of many beasts, including the European robin, the turtle and even the human. The Earth's magnetic field could affect chemical reactions involving these pigments, meaning that the turtle would see more or less what we see, but with a pattern of lights or colours on top that changes depending upon which direction the animal is facing. 'A bird facing toward magnetic north might see a big orb of light superimposed on its visual field and as it turns to the east that might become smaller or maybe split into two,' says Lohmann.

Iron man?

But what about us: can we sense magnetic fields? Back in the 1980s, Robin Baker of the University of Manchester, UK, blindfolded his students and drove them out into the countryside. When asked to point towards their campus, his test subjects did surprisingly well, unless Baker held a bar magnet to their temples. As a result, Baker wrote the book *Human Navigation and the Sixth Sense*. But almost no-one has managed to repeat these findings. 'There was a series of papers back and forth for about 10 years from 1980 to 1990 and then everyone grew tired of the controversy,' says Lohmann. 'As far as I'm aware there isn't anything new since that time.'

Lohmann reckons there's not yet any compelling evidence for humans having a magnetic sense. 'But Baker did make an interesting argument, which was that in our day-to-day lives the magnetic environment is severely disrupted,' he adds. 'We live in houses that often have steel beams, anything electrical makes electromagnetic fields. It's possible we have a magnetic sense but because we haven't grown up using it we may no longer be aware of it.'

This magnetic-field disruption could be upsetting turtles too. The steel in condominiums or hotels on the beach could throw off the magnetic field, as could coastal power lines or underwater cables carrying electricity from wind turbines. Even protecting turtle nests from invading raccoons with steel cages may disrupt the magnetic field that the eggs and hatchlings experience, making it harder to find their way back to that beach as adults.

Humans are affecting more than just the turtles' magnetic life. Those same hotels deter females from nesting, while the buildings' bright lights attract turtle hatchlings to scramble towards them, rather than down to the ocean: the youngsters mistake the artificial glow for the moon and stars reflecting off the water. What's more, the oceans are filling with plastic waste. To a turtle a floating carrier-bag looks like a jellyfish, one of its favourite foods. But a turtle with a stomach full of indigestible plastic is a turtle that will starve. The same goes for balloons; those charity balloon releases help worthy causes but they don't do turtles any good at all. And, as with many reptiles, the number of males and females that hatch from a batch of eggs depends on how hot they get. A loggerhead turtle nest at 28°C will hatch only males, one that's at 30°C will spawn a 50:50 mix of males and females, while one that's at 32°C will produce girl-power alone. That means climate change could make the male loggerhead turtle – and ultimately the species – extinct. Here high-rise hotels could, for once, assist by casting turtle nests into the shade and keeping them cooler.

Knowing about turtles' magnetic prowess may help to conserve these animals. So far no-one has managed to persuade other turtles, even animals of the same species, to nest on beaches where the original turtles are no more. The best known example is Bermuda. 'There were thousands upon thousands of green turtles nesting there in the 1600s,' says Lohmann. 'Because it was an easy source of meat, people harvested the turtles year after year and eventually that entire population was wiped out. For a couple of hundred years now, no green turtle has nested.' Transplanting green turtle hatchlings and eggs from Costa Rica didn't work – the turtles didn't return to Bermuda to nest. 'No-one was sure why,' says Lohmann. 'But it seems at least possible that reintroducing turtles hasn't been done in such a way that those turtles can imprint on the magnetic field of their new area. It's all theoretical at this point. But it could help.'

Meet the quantum mechanics

So loggerhead turtles have an in-built magnetic compass that lets them get back home even after a trip round the Atlantic. Thankfully we humans have something even better: in-car satellite-navigation systems, using signals sent to atomic clocks on satellites up in space. There's no chance of us getting lost (except if we ignore the spoken instructions) as we drive on a glorious summer's day to Mottisfont in Hampshire (see Chapter 4). We're off on a trip of our own to the *Furry Logic* summer picnic. Choosing a shady spot on the lawns overlooking the historic hall, we lay down a blanket, get out the cheese sandwiches, open a huge bag of crisps (we're so classy), pour some tea and unpack our pièce de résistance – home-made scones, clotted cream and strawberry jam. Perfect.

Unfortunately, we're not alone. Wasps have arrived. First one, then another, then a third, darting madly around us in search of a sugary treat. We try to shoo off the beasts but it's no use. More wasps turn up. One's crawling over the jam. Another's landed in the cream. The wasps are a complete (insert your own expletive) nuisance. Leaping up we tread backwards into a sandwich, knock over the tea and flail furiously about. With more wasps buzzing round our heads, it's time for plan B: shove everything back into the picnic hamper and dash for the car.

Wasps are one of the most unpopular animals on the planet. They have few fans and many enemies, but it turns out wasps (or at least some of them) are masters of electricity and expert at quantum mechanics. Before we explain how, let's make a case for their defence. First, without these yellow-and-black striped creatures, we'd be knee-deep in aphids and black fly. If you're a keen gardener, you can thank your local wasps for devouring these insects and keeping your cabbages in good nick. Second, many species are social creatures that live in giant colonies and have just one aim: to bring food back to their nests. They'll attack only if provoked or if they see a sudden movement, which is why swiping at one with a rolled-up newspaper is a bad idea. And here's a tip: if you're near a wasp's nest, stay still. Creating a disturbance encourages the wasps to rush out to see what's going on. If anything, wasps are more concerned about intruder wasps entering their colony. Should that happen, the inmates circle the outsider, before leaping on the enemy, chewing its wings off and stinging it to death. So it's not about you, it's them.

There are tens of thousands of species of wasps, but our focus is the Oriental hornet (*Vespa orientalis*). Found in many parts of north-east Africa, the Middle East and South-east Asia, it's mostly brown but has two yellow stripes on its head and two on its body – a warning sign for predators to stay away. It's a whopper too, roughly 2–3cm (0.8-1.2in) long. This wasp has made it into *Furry Logic* because it doesn't just eat sweet nectar and scavenge for insects, but also uses a branch of physics that grew from studies of electricity and magnetism. We're talking about quantum mechanics.

Quantum wonderland

Ordinary mechanics (minus the quantum) is all about big things in the everyday world like the mosquitoes and mantis shrimps we met in Chapter 2, and how they move or change under the action of forces. Quantum mechanics, on the other hand, describes the world at the level of atoms and molecules. And it's weird. Developed in the early twentieth century by top physicists such as Niels Bohr, Paul Dirac, Albert Einstein, Werner Heisenberg and Erwin Schrödinger, the theory lies at the heart of many electronic devices – computers, lasers, smartphones, you name 'em. Without quantum mechanics, modern life would be rubbish (try living without your mobile for a week and you'll see).

The quantum world is counter-intuitive. An electron, for example, spins in many different directions at the same time. But if you try to measure which way it's turning, the electron will seem to have been spinning in one direction only. And just as you can't stand on a ladder at any point you like but only on the rungs, so an electron can't have any energy, but is restricted to certain values. Another oddity is that you can't measure certain combinations of physical quantities, such as a particle's position and momentum, at the same time with complete accuracy, no matter how advanced your equipment. Narrow down where an electron is, say, and you'll have little idea about its momentum; pin down its momentum and you'll be mostly clueless as to where the electron is. Known as Heisenberg's uncertainty principle, this tenet of quantum mechanics is the basis of that very rare beast: a funny physics joke. Heisenberg gets caught speeding by a traffic cop, who asks him if he knows how fast he was going. 'No,' replies the great man, 'but at least I know where I am.' It's OK if you didn't laugh ... the bar for physics humour is low. Two atoms bump into each other. 'I think I've lost an electron!' says one. 'Are you sure?' replies the other. 'Yes, I'm positive!'

While we're still grappling with the weirdness of quantum mechanics, the Oriental hornet got there first. It's been putting quantum principles to use for far longer than us. To find out how and why, we need to pry into this wasp's domestic arrangements. These insects live in colonies, each dominated by a queen wasp, who mates with one or more males in summer and stores their sperm over the winter months. When the queen emerges from her hibernation in early summer, she creates a nest of paper-thin combs. There she lays eggs that become sterile female workers. These offspring dutifully serve their queen by clubbing together to build a colony around her so she has room to lay hundreds and even thousands more eggs. They'll also forage for food, guard the nest and take care of the brood.

Come late summer – and with one eye on her legacy – the queen lays eggs that develop into fertile females and males, the guys buzzing off to mate with a young female from another colony. That female becomes a new queen, keeping the wasp dynasty alive. It's a collective achievement that has fascinated researchers all the way back to the fourth century BC when Greek philosopher Aristotle noticed that hornets don't just feed on nectar. They also kill insects, which they chew and turn into a slurry that they feed to larvae developing in the nest.

The good news regarding Oriental hornets is that they won't nest in your house. Nor will you find them living in trees or shrubs. Instead, these wasps hang out in intricate underground burrows, which armies of workers hollow out by digging with their mandibles. Carrying the soil in their mouths, the workers head to the nest exit before flying out about 10m (about 30ft) from home. After dumping the soil in mid-air (the naughty litter bugs) they return home for more digging.

Most species of hornet fly in the early morning, restricting their flight times to stay cool and avoid the potential damage to their bodies from the intense ultraviolet light that shines down at hotter times of day. But in 1967 Jacob Ishay from Tel Aviv University in Israel was counting how often Oriental hornets leave their nest to drop off soil when he noticed that these insects don't concentrate their activity in the early morning. Like mad dogs and Englishmen, they love nothing better than the heat of the midday sun. The more ultraviolet light there is, the more soil they dig and the more they leave their nest. These sun-worshippers can be up to a hundred times more active at noon than in the morning.

Ishay assumed these hornets must love the sun because they somehow harness sunlight. For many years that idea was idle speculation and Ishay became sidetracked by another oddity of Oriental hornets: in their nests, they always create their combs by starting with a small stem and working downwards. To explore why, in 1992 Ishay sent more than 200 Oriental hornets into orbit on a NASA Space Shuttle. There's little downwards gravitational pull on the Shuttle and Ishay wanted to see if Oriental hornets behave differently in zero-gravity conditions from how they do on Earth. He even hoped the work might help explain whether astronauts feel sick in space because the lack of gravity disorientates them. Sadly for astronauts – but good news for any wasp-haters – the experiment's water system failed and almost all Ishay's insects died before he got any meaningful results.

Surface thinking

Enough of Oriental hornets in space. Ishay's hypothesis that these animals exploit sunlight finally came to the fore – but sadly only after his death in 2009. It was then that a former student, Marian Plotkin, took a close look at the Oriental hornet's hard outer surface. Rather than using a conventional light microscope, which can show only so much detail, Plotkin and some of his colleagues studied the hornet with an atomic-force microscope. At the heart of this kit is a small pointed tip that you scan over the surface of an object, like reading Braille with your fingers. The tip is connected to a cantilever beam, which moves a little or a lot depending on the size of the force between the tip and the atoms below it. By recording the force at many points on the surface of the hornet, Plotkin found something strange.

Earlier we mentioned that the outer skin, or cuticle, of an Oriental hornet has two parts: brown and yellow. Plotkin found that the two areas are very different. The brown bit – so coloured because it contains melanin pigments like those in our skin – has a surface that isn't flat, but looks like a corrugated roof or a ridge-cut potato crisp. This up-down-up-down surface means that sunlight is more likely to be trapped in the hornet's skin than to bounce off. Below its ridges, the hornet's cuticle has about 30 layers containing rods less than a thousandth of a millimetre in size that trap light even further. Plotkin reckons that the ridges and the layers together help the Oriental hornet to squirrel away 99 per cent more light than if the brown surface was flat.

As for the yellow part of the Oriental hornet, Plotkin's microscope revealed that its surface is more undulating and contains joined-up oval blobs each pockmarked with at least one small pit. We don't know the point of the pits, but this yellow part of the wasp's cuticle is as good at trapping sunlight as the neighbouring brown area. The yellow colour, Plotkin discovered, comes from a layer beneath the surface consisting mostly of barrel-shaped granules containing xanthopterin, a pigment that's also found in butterfly wings and human urine. This molecule's a cracker at absorbing sunlight, and Plotkin wondered if the Oriental hornet uses it to convert light into electricity. Could this animal, in other words, be a living, breathing solar cell?

To understand how a molecule can turn light into electricity, we need to return to an oddity of the quantum world – that electrons can only have certain energies. If those energies were the rungs on a ladder, an electron can sit on one rung or another, not in between. The electron can, however, jump from a lower rung to a higher rung if it's pushed by light with enough energy to help the particle bridge the gap. A xanthopterin molecule would need to absorb ultraviolet light with a wavelength less than 386 nanometres for this electron transition to occur (although we're not sure which electron in the molecule does the jumping). And once you've set electrons moving, you're well on the way to generating an electric current.

To prove whether xanthopterin can turn sunbeams into electricity, Plotkin just needed to wire up this pigment into an electrical circuit. If the energized electrons could be made to flow to an electrode, he'd have electricity. And if humans can make a current from this molecule, you can be pretty sure the Oriental hornet does it too. All it needs is enough xanthopterin and enough light.

In 2010 Plotkin made an artificial solar cell by sandwiching a film of xanthopterin between two glass electrodes, one coated with titanium dioxide, a semiconductor that's also used in paint and sun cream. Shining a lamp onto his device, he found that roughly three in every 1,000 photons of light boosted an electron in the xanthopterin to a higher level, enabling them to move through the semiconductor to the plain glass electrode. Converting 0.3 per cent of light into electricity isn't a huge figure – the solar cell powering your garden light bulbs, which has silicon as its key ingredient, will typically turn about 10 per cent of photons into electrons. But Plotkin's achievement was a breakthrough nonetheless. 'No-one thought xanthopterin was used to convert sunlight into electricity,' says Plotkin. 'We've known for decades that certain photosynthetic bacteria and algae can turn light into chemical energy to fuel themselves. Now for the first time we've shown animals can convert light into other forms of energy too.'

Plotkin thinks that the Oriental hornet could also use its brown melanin pigment as a solar cell, though that's one experiment he hasn't tried yet. 'The hornets probably combine the melanin and xanthopterin properties in this light conversion process, although the full effect is currently unknown,' he says. Quite why the Oriental hornet has an in-built solar cell, however, remains an enigma. The current may give the hornet an extra power source when it flies. Or it may use the electricity to keep cool: insects don't have sweat glands and can overheat.

Anyone thinking of finding out does need to be careful. Experimenting with Oriental hornets is difficult and dangerous; Plotkin got stung four times. 'It's horrible,' he recalls. 'The venom's really potent and it took me several days each time to recover.' Surprisingly, though, the more he was stung, the better Plotkin felt. 'This isn't a particularly scientific analysis, but during the four or five years I was working with Oriental hornets, I never got sick once.'

When two become one

If there's one thing we've learned in this chapter, it's that animals have been harnessing natural phenomena long before we humans got there. Electric eels stun prey by firing out volleys of high-voltage pulses just like a police Taser. Bees were picking up electric charge well before the ancient Greeks sussed that trick by rubbing fur and amber together, and loggerhead turtles navigate with their own compass that picks up the Earth's magnetic field. We ended by discovering how the Oriental hornet has an in-built solar cell that creates electricity from sunlight by exploiting the weird world of quantum mechanics.

There was something more fundamental to this chapter too. As Faraday, Ampère and other scientists realised in the early nineteenth century, electricity and magnetism are two sides of the same coin. But it was Scottish physicist James Clerk Maxwell (1831–79) in the 1860s who brought the two phenomena under one roof. To use physicists' lingo, he united the electrical and magnetic forces to create one: the electromagnetic force.

This brings us neatly to the final physics topic in *Furry Logic*. As Maxwell was astonished to find, when electric and magnetic fields travel through space together, they create something without which life on Earth would not exist. No plants. No Oriental hornets. No Toby (Maxwell's beloved pet dog, which adorns his statue on George Street in Edinburgh).

That something is: light.

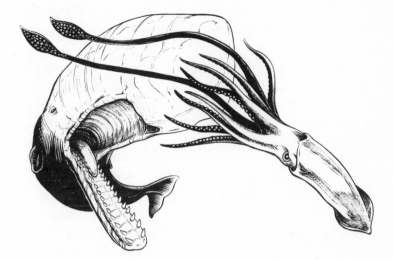

CHAPTER SIX

Light: A Final Physics Fandango

ANTS AND BEES THAT USE A TRICK OF THE LIGHT * FLASHY
CUCKOOS * DEADLY SPITTING FISH * UNDERWATER
CHAMELEONS * THE SQUID WITH THE GIANT EYES

Getting antsy

Head to a supermarket and the first thing to greet you –
after the suspicious security guard – is a warm waft of sweet
air as if from the in-store bakery. By pumping out this
scent, store bosses lure you towards the almond croissants,
crusty cob loaves and cartwheel-sized cookies at the back of
the shop, tempting you to impulse-buy as you trek past the
other stock. Apart from using money and loyalty cards,

we're no different in this respect from many species of ant, which 'sniff' out food with their antennae. The insects track trails of pheromones – chemicals laid by other ants – to get from their nest to a food source and back.

In the deserts of North Africa, however, it's a different story. Food is scarce and pheromone trails fizzle out fast in the heat. That's if the ants lay a trail at all; some species such as *Cataglyphis* desert ants don't. In their hunt for dead insects or spiders, these ants scuttle this way and that. Plot their path and it's hugely inefficient, like a straggly ball of string. Some individuals meander more than 100m (330ft) from their nest. But these black-bodied ants, which weigh barely 10mg (0.0004oz) and are about 6mm (0.2in) long, are nowhere near as unfocused as their trails suggest. When they've found food, the insects stop wandering about. This time their route is taut – they make a beeline, or rather an antline, straight back home. On their outward journey, the ants somehow tracked how far, and in which direction, they wandered. It's as if they have an in-built measuring tape and compass so they can work out the quickest way back.

Discovering how the ants achieve this feat will take us on a roundabout journey of our own, from Tunisia to Germany, then on to Austria, the UK and back to Germany. Researchers could have solved this quest much faster if only the first expert had had the right kit. Instead it took more than 80 years and the help of another insect: the bee. Getting from A to B on this trail of discovery is a case of going from ant to bee and back again, but not before the bees have led us a merry dance.

The ant doctor

First to study the f-ant-astic homing ability of desert ants was maverick medic Felix Santschi (1872–1940), who was a dab hand at cutting cataracts from patients' eyes with a copper knife. Since he didn't have a high-school diploma, Santschi couldn't practise in his native Switzerland. So in 1902 he moved to the ancient North African city of Kairouan in present-day Tunisia. There he was medic by morning and evening, ant researcher by afternoon. Altogether he published more than 200 scientific papers describing nearly 2,000 different species, subspecies, varieties and sub-varieties of ant. Living far from most other scientists and submitting much of his research to obscure French-language journals, Santschi stayed under the radar until long after his death. Which is a shame, as his findings could have put us on a more direct route to ant understanding.

In a photo on the AntWiki website, Santschi holds the iron-ring handle of an ancient door patterned with metal studs. He's wearing a dark suit jacket with an open-necked white shirt and, in true doctor style, a pen pokes out of his top pocket. His goatee is greying and his eyes look as if they've squinted at too many small insects. After years of study, Santschi didn't believe desert ants follow chemical trails. If they did, why would they go round the houses on their way out but come straight home? He figured that instead the ants must note something about the world around them. Not physical objects such as a leaf, a stone or – like Hansel and Gretel – a trail of breadcrumbs; these might blow away, or be moved or eaten. Santschi suspected ants use the one thing that's everywhere in the desert: the Sun.

In 1911 Santschi tested his idea near Kairouan by showing ants the Sun in a mirror, so that it appeared to be in the wrong place in the sky. Confused, many species swiftly changed course. Those who had been returning to their nest headed away from home. The ants, it seemed, used the Sun as a guide. To the few scientists who got wind of it, this finding was controversial. The Sun moves across the sky, so why rely on it rather than a local landmark? Surely ants don't have to be astronomers to find their way home?

Stranger still, some species, notably the Sahara Desert ant (*Cataglyphis bicolor*), didn't change course when Santschi 'moved' the Sun but carried on heading back to the nest. These ants weren't using the Sun as a landmark; they were exploiting its light in some other way. Even if Santschi's assistant restricted their vision to a circular patch of sky by moving a cardboard cylinder, about 50cm (20in) in diameter and 25cm (10in) high, along with them as they walked, these insects still navigated correctly. Despite not seeing the mirrored Sun directly, these ants reached their goal.

As Santschi asked in a 1923 paper: 'What is it in this small patch of blue sky that guides the ant back home?' The doctor reckoned ants navigate using variations in the intensity of sunlight, probably at ultraviolet (UV) frequencies that humans can't see. It was a decent assumption with just one snag: it was wrong. Santschi could have tested his idea by exposing ants to UV light from a mercury lamp. But there weren't many lamp stores in the middle of Tunisia, so there ended his contribution to ant navigation. When Santschi died in 1940, few scientists were aware he'd shown that ants use light from the Sun, as well as the star's position, to find their way. Locals in Kairouan, on the other hand, admired Santschi, his twin efforts in medicine and zoology earning him the nickname *Tabib-en-Neml*, the 'ant doctor'.

From ants to bees

Out in the Tunisian desert, Santschi had come tantalisingly close to discovering how ants direction-find. But it took another scientist, working on a different insect, to come up with the answer: Austrian-born zoologist Karl von Frisch (1886–1982). Clearly an animal lover, in one photo von Frisch sits in suit and tie, a floppy-eared dog blurring as it bounds onto his lap for a cuddle. In other pictures, he's less formal, wearing the white shirt and embroidered braces of traditional Austrian costume. Based for much of his life at the University of Munich in Germany, von Frisch discovered that European honeybees (*Apis mellifera*) dance to tell their hivemates where to find pollen and nectar. Inside this dance lay the secret to bee navigation – and that of ants too. Bees, it turns out, gather direction data just like ants.

If a worker bee finds food up to about 50m (165ft) from the hive, von Frisch revealed, on her return home she loops round on the honeycomb in a small circle a couple of times then switches direction. The bee does this back-and-forth 'round dance', like the big hand on a confused watch, for several minutes. Other bees follow her path before flying off to scout for the nectar. The round dance says 'there's food not far from our hive and it smells like the food sticking to me but I'm not telling you which direction it's in, just go and have a look'.

A bee passing on the news that there's food further away, up to 15km (9 miles) from home, does a more complicated dance. This time, the honeybee crawls in a straight line, loops in a semicircle back to the start, crawls the straight stretch again then loops another semicircle, this time in the opposite direction. Her path resembles two capital Ds back-to-back. On the straight run, the bee shakes her body from side to side in a 'tail-wagging' dance. To tell other bees how distant the food is, the dancer adjusts the length and speed of each circuit. The code is simple: the further away the nosh, the fewer complete waggle dances the bee does. If dinner is, say, 100m (330ft) away, the worker does 10 short laps in 15 seconds, but if the food is 3,000m (9,900ft) away, she dances three much longer circuits in that time. The other bees watch carefully; sometimes they'll monitor several different dancers and take an average for the distance to the food, in a form of bee crowdsourcing. This link between distance and the number of waggle dances is so exact, von Frisch found, that he could time the dances with a stop-watch to work out how far off the food was. Now that's a case of furry logic.

To know how far she travelled to fetch her food, the dancing bee looks at how much energy she burnt during the flight: she uses a fuel gauge. Telling other bees the distance to the food isn't much use, however, unless the dance can also explain which *direction* to fly. If you ask a visiting friend to buy a loaf of bread from the bakery a mile down the road, chances are your pal will wander the streets empty-handed unless you also indicate the direction. So how does a bee work out directions? The great minds of von Frisch and Santschi were studying the same question, even if von Frisch only got wind of Santschi's work in 1948, some eight years after Santschi's death.

By then von Frisch had moved back to Austria, after Second World War bombing destroyed the Zoological Institute in Munich. At the University of Graz he looked at bees dancing on a honeycomb, the wax mesh of hexagonal cells where these insects store their honey and lay their eggs. Von Frisch found that just as a bee has a secret code to tell hivemates how far off food is, she can also pass on the direction to fly. A beehive contains many honeycombs standing vertically, and he discovered that the angle of a bee's waggle run, relative to the vertical, indicates the angle between food, hive and the Sun. If the hive, food and Sun all lie in a straight line, the bee does its waggle run vertically upwards. But if, say, the line from food to hive lies at 60° anticlockwise to the line from hive to Sun, the straight stretch of the bee's waggle dance runs at 60° anticlockwise to the vertical. In other words, the bee uses the vertical, which she detects through feeling the pull of gravity on her body, to symbolise the position of the Sun. And, in her direction find and tell, she relates the position of the food to this vertical, and hence to the Sun. Through the waggle dance, a colony of honeybees can harvest a new source of food without individuals wasting time struggling to find it.

Place a honeycomb horizontally in a dark, windowless hive, and the bees are confused. No longer able to use the

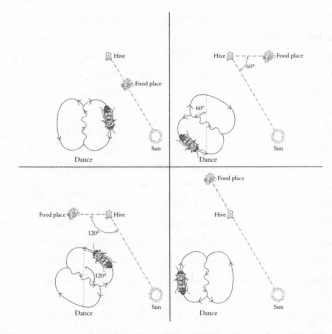

Figure 6.1 **Special path.** *Honeybees tell their hivemates where food is using a waggle dance, with the angle that the straight part of the dance makes with the vertical being linked to the angle between the Sun, the hive and the food.*

vertical as a reference direction, their dancing patterns become a mess. Shine a strong electric light into the hive, though, and the bees are back in business: the straight run of their waggle dance relative to the lamp is at the same angle that the path from hive to food source makes with the Sun. The bees, in other words, use the lamp as a reference point. Now here's the crunch. Do bees need to see the Sun itself or, like Santschi's ants, is a patch of Sun-free sky enough? If it's cloudy or the Sun's behind a mountain or a tree, can bees still figure out where they buzzed off to for their food? In the summer of 1948, von Frisch found out by allowing bees on a horizontal honeycomb to see a small area of blue sky through a window he opened up in the hive (just as Santschi had used a cardboard cylinder to let ants see a

Sun-free patch of sky in 1920s Tunisia). Even if a returning bee could see barely a 10° span of the sky from inside (90° being the angle between the vertical and the horizontal), its waggle dance still informed fellow bees of the direction of their food.

Blue-sky thinking

There must be more to the sky than humans can see. Let's investigate. For a start, why is the sky blue? Luckily we've begun with an easy one. Molecules and particles such as dust or smoke scatter sunlight that has a high frequency more than they scatter light of low frequency. A lot more: the amount of scattered light increases with its frequency to the power of four. So air scatters blue light (short wavelength but high frequency) about 10 times more strongly than red. This is also why fine smoke and mist look blue. What's more, this scattered light is polarised.

Whoa there. 'Polarised'? What's that? Time to rope in more string. Not straggling this time, like the ants' trails, but a length of rope held between two people and waggled about. But first, let's head back to the nineteenth century when Scottish physicist James Clerk Maxwell calculated, to his surprise, that magnetic and electric fields move at the speed of light. We can think of light as a set of waves made up of electric and magnetic fields, he realised – it's an electromagnetic wave. Physicists love Maxwell's equations so much that they came joint top in the *Physics World* all-time favourite equation poll we mentioned in Chapter 2. As we learned earlier, F=ma (Newton's second law) came third. One point to Maxwell, who gets a bonus point because, when told 'You have done great things but you stand on Newton's shoulders,' physics genius Albert Einstein replied, 'No, I stand on Maxwell's shoulders.' Shoulders were a theme: centuries earlier Newton, writing to rival scientist Robert Hooke, had said of his optics work, 'If I have seen further, it is by standing on the shoulders of giants.'

Enough of this physics acrobatics. The values of light's electric and magnetic fields fluctuate up and down along axes at right angles to each other. If the electric field was the rope we mentioned wiggling left and right, the magnetic field would be another rope wiggling up and down. For unpolarised light, the electric field is always changing, like a rope wiggling first horizontally, then at an angle to the horizontal, then at a different angle and so on, changing direction a hundred million million times a second. As does the magnetic field, which is at right angles to the electric field. For polarised light, in contrast, the electric field oscillates in the same direction all the time. If the electric field were a rope waggling vertically up and down, we call the light vertically polarised; waggling horizontally, it's horizontally polarised.

Light relief

Light from the Sun arrives at the top of Earth's atmosphere in unpolarised form. As it travels onwards, molecules in the air scatter the light, polarising it by different amounts and in different directions according to the angle they scatter it through, giving the sky a distinct 'polarisation pattern'. Although our eyes can almost never tell polarised light from unpolarised, we have cameras that can reveal how much the light is polarised – and in which direction – at every point in the sky. If the Sun is directly overhead, as on the equator at noon, the direction of polarisation points in a series of concentric rings around the Sun; at each point on a ring the polarisation is at a tangent to the circumference. Go round the circle and the polarisation direction changes until you get back to the start. As the Sun sets in the west, a band of very strong polarisation runs from the north, up across the highest point in the sky and down to the south.

After finding out about polarisation from his physics pal Hans Benndorf (1870–1953), von Frisch wondered if that's how honeybees can waggle-dance on a horizontal honeycomb in a hive with only a small patch of sky on show. Do they, in other words, orient themselves using the sky's polarisation pattern when they can't see the Sun itself? To find out, von Frisch put a sheet of Polaroid in the window of his hive above a bee dancing on a horizontal honeycomb. Polaroid is a polymer, like polythene or polystyrene, with long spaghetti-like molecules that all line up in the same direction. Developed by co-founder of the Polaroid Corporation Edwin Land in the 1930s, this plastic absorbs light that's polarised in the direction of the molecules but lets light polarised in other directions pass through. As a result, the material acts as a polariser. When von Frisch rotated the Polaroid sheet to change the direction of polarisation of the light reaching the comb, the bees altered the direction of their waggle dance. QED bee.

There's a rumour that Vikings used polarisation too, navigating their longships before the invention of the compass by looking at the sky through a sunstone, a piece of crystal that bends light through different amounts according to its polarisation, creating two separate images. But this might only be as true as the 'fact' that Viking seafarers wore horned hats (they didn't). On the bee front there's more to learn as well: we don't fully know how they detect light's polarisation, for a start. As von Frisch put it in his 1950 book *Bees: Their Vision, Chemical Senses and Language*, 'the bee's life is like a magic well: the more you draw from it, the more it fills with water.'

Back to bee-sics

Von Frisch's discovery that bees use polarised light to navigate encouraged David Vowles from the University of Oxford, UK, to extend Santschi's ant studies. One sunny evening in 1950 Vowles ventured out to the flat meadows next to the River Cherwell for a spot of ant-watching. By restricting the insects' vision with a tube, he showed that these *Myrmica lavinodis* ants in the Thames Valley could find their way home with the aid of only a patch of Sun-free sky, confirming Santschi's work with Tunisian ants. What Vowles did next clinched it. Back in the lab, he shone a beam of unpolarised light through a sheet of glass onto a *Myrmica ruginodis* ant on a horizontal surface. The ant wandered, rarely going in one direction for more than 4cm (1.5in). When Vowles passed light through a Polaroid sheet, however, the ant would eventually set off in a straight line for up to 20cm (8in). If, after the ant had moved 5cm (2in), Vowles rotated the Polaroid clockwise by 30° about the vertical axis, the ant veered off in a direction 30° clockwise to its original route. The results were similar for other rotations. Like bees, these ants use polarised light to orient themselves. And so, around 30 years after Santschi's studies, the secret of ant navigation had finally been cracked. It's time for the final leg of our journey.

Ants on stilts

The Sun's polarisation pattern only tells ants which *direction* they've gone: north, south, east or west. To calculate the quickest route home, an ant also needs to know *how far* it's strayed along each compass bearing. How ants do that remained unsolved until 2006, more than 60 years after Santschi's death and 50 years after Vowles's experiments. That was when Matthias Wittlinger and Harald Wolf from the University of Ulm in Germany, along with Rüdiger Wehner from the University of Zurich, Switzerland, did the legwork to solve the final part of this riddle.

To mimic the desert, Wittlinger and his team built a 10m (33ft)-long metal channel, painted the insides grey and poured in a layer of fine grey sand. At one end they put a nest of *Cataglyphis fortis* ants, at the other a container of biscuit crumbs. Once the ants discovered the food, the team caught the animals and placed them – biscuit crumb in mouth – at the end of another channel parallel to the first.

This second channel was identical except that it was 20m (66ft) longer and didn't have a nest at the far end. Unaware, the ants headed down the imposter channel. After roughly 10m (33ft) they paced back and forth, searching where they thought the entrance to their nest should be. Somehow they had a sense of how far they needed to go to get home, even after a scientist had tricked them.

Ants may do this in two ways, Wittlinger and co reasoned. Either they have an internal clock that tells them how long it takes to walk from nest to food – as long as they walk at the same pace on the way back, they just need to time their return journey. Or the ants count their steps and assess their stride length on the way out, then work out how many strides of a certain length they should take on the way back. To find out whether timing or counting is the secret, the researchers glued a pig bristle to each leg of ants that had made it to the biscuit crumbs. The plan was that these six 'stilts', which lengthened each leg by about 1mm (0.04in), would give the ants a bigger stride. If ants count their strides, then with longer legs they'd overestimate the distance to their nest. 'The most challenging [part] of the experiment was not only to glue on the stilts but actually to keep the ants "happy" and motivated enough to take a food crumb after the preparation,' says Wittlinger. 'Only ants that carry a food item can be tested since only then do we know about their motivational state for sure.'

When stilted ants entered the longer, second channel, they duly overshot their now invisible nest by a little over 5m (16ft). This proved that they were assessing their steps to measure distance but hadn't compensated for their newly enhanced stride. To double-check, the team put these ants on stilts back in their nest for a couple of days, so they could walk to the food and back along the original channel many times. When rudely interrupted from their dinner and plonked in the second channel once again, the ants looked for their nest after about 10.5m (34ft). They had learned how many steps they needed with their modified limbs. The distance wasn't spot on – the nest was 10.2m (33ft) away from the crumbs, not 10.5m (34ft), a discrepancy the team put down to ants on stilts moving at a slightly different speed. As Wittlinger realised, ants can't just rely on counting their steps. Life's complicated and they must account for stride length too. 'Stride lengths vary with varying walking speeds,' Wittlinger points out. 'The speeds to the food and back home usually vary so the numbers of steps are different.'

But the bottom line was clear. Ants don't have an internal clock but an in-built pedometer, which they combine with polarisation detection to work out how far and where they've wandered to find food. Then, somehow, they calculate the most direct route back. This was one small, but slightly bigger than usual, step for an ant and a giant leap for mankind's understanding of animal navigation. And it brings us full circle, back to Santschi's original antics in the Tunisian desert. If only he'd had a mercury lamp, perhaps researchers could have taken a direct route to discovering ants' navigation secrets. Instead, they had to scrabble about for answers like the insects looking for food on their outward trip.

Ants and bees were just the start of a much longer journey. Today we know that flies, spiders and beetles detect polarised light too, as do other insects and a heap of aquatic animals, from octopus and cuttlefish to mantis shrimp. Animals may use polarised light for everything from navigating their world and spotting predators to camouflaging themselves and communicating. We don't yet know for sure how many animals can – or can't – see polarised light. In the words of Shelby Temple of the University of Bristol, UK, whom we'll meet soon: 'In terms of trying to understand polarisation vision, we're like children in kindergarten trying to learn ancient Greek.'

A spot of colour

We may be toddling round the edges of polarisation vision, but we've had much longer to study colour. In his 1704 book *Opticks*, Newton – it's time we brought him back without saying he was wrong or insinuating that Einstein preferred Maxwell – pondered whether a peacock's colours arise from the way the light bounces off the feathers rather than being due to absorption by pigments. That said, it wasn't until the late twentieth century that physicists began to exploit this phenomenon.

Animals use colour to find food, attract mates, hide from or scare attackers, communicate their state of health or, in the case of some colourfully bottomed primates, fertility, and more. In physics terms, colour is the wavelength of the pulsating electric and magnetic fields that make up the light wave. In our rope analogy, it's the distance for the rope to complete one wiggle up, down and back again. Humans can see light with wavelengths of 400 nanometres (violet) to 700 nanometres (red). As Newton discovered, by shining light into a prism, you can split it into its seven constituent colours – red, orange, yellow, green, blue, indigo, violet – since the glass bends each wavelength by a slightly different amount. Before this, people thought colour wasn't a property of light but was somehow generated by the glass or other object they were looking at. Newton's commitment to vision research was extreme; he stuck things down the side of his eyes to see how squashing his eyeballs altered his sight. It produced colours, but DO NOT TRY THIS AT HOME. Unlike his 1687 classic *Principia Mathematica*, *Opticks* was published in English and seems to have been an easier read, or at least no-one's on record offering a prize to people who can explain it.

Talking of prizes, as well as beating Newton in the favourite equation stakes, Maxwell was the one to create the first colour photo. In 1861, this Scottish laird took three separate black and white photos of a tartan ribbon illuminated in turn by blue, red or green light. Then he projected all three photos on top of each other, using a different colour of light for each. This recreated an image of the ribbon in glorious colour. And it was Maxwell who realised that there are more wavelengths of electromagnetic radiation than we can see – his four equations of 1868 predicted the existence of radio waves, X-rays and more. Beyond the visible electromagnetic radiation that we call light lie a host of other beams. Increase the wavelength of visible light and you hit, as we discussed in Chapter 1, infrared, then microwaves, which have a wavelength of about 1mm (0.04in) to 1m (39in), and radio waves, which have wavelengths up to 100km (62 miles). All these waves travel at the speed of light so as you increase the wavelength, the frequency with which they pulse decreases, making the wave less energetic. Increase the frequency to head down the wavelength scale, and past the visible you get into ultraviolet (UV) light, then X-rays (wavelength 0.01–10 nanometres) and finally gamma-radiation (wavelength less than 0.01 nanometres), as given off by some radioactive materials. Chapter 1 described a few of the animals that detect heat in the form of infrared radiation, with the fire beetle topping the sensitivity list. Our own eyes tell us we can sense visible light, if the clue in the name isn't enough. But some animals see light past the end of the visible spectrum too – in the ultraviolet.

Bird's eye view

Haggard and exhausted, her pale brown feathers bedraggled since she barely has time to eat, let alone preen, a female dunnock (*Prunella modularis*) returns to her nest of twigs and moss. Trapped between the two halves of her beak is a small beetle. Facing the sparrow-sized bird are five hungry mouths, gaping wide to reveal bright red-orange skin. Which to choose? It's a tough decision. The chicks that get most food will thrive. But they must compete with their siblings for the limited supplies. One of the youngsters' gapes is a little bigger and brighter red than the others. And that's the one the adult plumps for, popping the beetle inside for the chick to gulp it down. It's a wise choice because a bright mouth may signal that a chick is healthy – it's an 'honest indicator', as the bird must put extra energy into creating bright pigments, showing that it's doing well enough to have resources to spare. There's no way to cheat and make a strong colour at no cost. Alternatively, a bright red gape may show that the chick is hungry. A well-fed baby needs plenty of blood in its stomach to help it digest, but a bird with an empty belly can divert blood flow to its mouth. So a red mouth could be an honest signal that the chick needs food.

Nearby, another dunnock nest has only one chick. But it's enormous and its gaping mouth is much brighter red than usual. For some reason, at this nest the female feeds her freakishly large chick as often as she can. She may even put more effort into finding food for her baby than her neighbour with five offspring does. After a while, the hungry youngster grows much larger than its parents, becoming nearly 30cm (12in) long. It's way too big for the nest – its head and tail stick out either side. As you may have guessed, this chick is no dunnock: it's a common cuckoo (*Cuculus canorus*).

Nest egg

This cuckoo is not alone in its freeloading. About 1 per cent of bird species act this way – they're 'obligate brood parasites'. Laying your egg in another bird's nest lets you access the world's ultimate crèche: there's 24/7 care, no waiting list, and you never have to pay the bill. It's the bird equivalent of throwing a baby out of a high chair, plonking your own infant in its place, then heading to the pub. For ever. A female common cuckoo chucks one egg out of her chosen host's nest before laying her own egg in its place. If the cuckoo egg hatches first, the early-bird chick pushes its rival eggs out of the nest. Now it can catch the worm – it's won the undivided attention of its new foster-parents. And if the cuckoo emerges after the host bird's chicks, it shoves its step-brothers and sisters over the edge to their death. It's the story of *Cinderella*, only an Ugly Sister wins and there's no Fairy Godmother.

Even when the cuckoo chick has disposed of its rivals, its troubles continue. The baby bird needs more food than its foster-parents expect it to; they think it's a dunnock. The extra-bright, and larger, mouth of an imposter cuckoo chick, biologists suspected, is a 'supernormal stimulus' aimed at making 'parents' bring the additional food the chick requires to grow into a 30cm (12in) adult cuckoo, rather than a sparrow-sized dunnock.

There are supernormal stimuli elsewhere too – female oystercatchers (*Haematopus ostralegus*), for example, prefer to incubate abnormally large eggs, and many humans (*Homo sapiens*) are attracted to surgically enhanced, freakishly large breasts. Marketing execs, meanwhile, create ads that are larger than life, brightly coloured and loud, to tempt us to buy more. The term applies to any extra-strong, extra-bright or extra-loud signals designed to make another animal exaggerate its response. For a stimulus to reach supernormal status, it must both be a stronger signal than usual *and* trigger an enhanced reaction. Superman wouldn't be super if he hid in his bedroom. So does a cuckoo chick's gape tick both boxes and reach the dizzy heights of a supernormal designation?

Various biologists have investigated by painting the throats of baby host birds such as dunnocks, robins and reed warblers with red dye. Even though they had redder gapes than usual, the doctored birds didn't receive more food from their parents. Their painted throats failed the supernormal test as they didn't win a greater response. So why do cuckoo chicks spend energy making their pigments bright if they don't gain any extra dinner? Let's turn to Martin Stevens of the University of Exeter, UK, who started working on bird vision because he was fascinated by the camouflage and warning signs used by moths and caterpillars. Since these invertebrates are generally hiding from or trying to scare away birds, it's vital to know what the birds see. Stevens had his own suspicions about the throat-dying. 'Their manipulations when they changed the mouth colours were not based on how birds see things but on how humans see things,' Stevens says of these throat-painting researchers. 'So it's a little difficult to conclude anything about what that means.' It took a game of bird hide-and-seek in the mountains of Japan to test Stevens's thinking.

Bird-brained

It's hard for vision researchers to picture what birds see. Avian eyes don't work like ours. Many birds can perceive UV light, which has wavelengths less than 400 nanometres. This was 'convincingly demonstrated' in the 1970s, according to Stevens. Many other animals, including some species of insect, reptile, amphibian and fish, see UV too. We've known that ant vision works in the UV since a discovery in the 1890s by banker, scientist and Liberal politician John Lubbock (1834–1913) – an unusual job combination that earned him a depiction as a flying insect in a *Punch* cartoon in 1882, accompanied by the text 'How doth the banking busy bee, improve his shining hours, by studying on bank holidays, strange insects and wild flowers!' The ant world, as we saw with Santschi, attracts enthusiasts with non-ant day jobs. Lubbock was a neighbour of Charles Darwin; Downe, in Kent, wasn't a great place to be a small animal – your chances of being collected by a naturalist were way above average.

Naturalists, and other humans, have three types of cone cell, photoreceptors in the light-sensitive retina at the back of our eyeballs. Each contains a pigment that's most responsive to red light (long wavelengths), green light (medium wavelengths) or blue light (short wavelengths). Birds go one better. They have red, green and blue cones but also, depending on the species, a cone that responds either to violet light and a little ultraviolet, or to ultraviolet. They're tetrachromatic to our trichromatic. Some humans may be tetrachromats too, and able to see more colours than the rest of us. But it's not clear how these lucky individuals' extra cones are wired up. Generally, human cones can detect a little UV but they're not very sensitive. What's more, the lenses in our eyes block most UV light from reaching our retinas. So we miss out on a multitude of flower patterns and animal markings that are out of our league.

Our misguided assumption that birds see like humans has held up our understanding of their biology. Instead of using their own senses, vision researchers must turn to technology to get the full picture. 'There's a whole part of their [birds'] world that we miss out on,' says Stevens. 'You need a lot of specialist equipment to understand what sort of visual information is available to a bird. It makes it more interesting and more challenging.'

Watch the birdie

Back to the original question: do cuckoo chicks use supernormal stimuli? To find out, Stevens headed to the slopes of Mount Fuji in Japan to assist in a quest for the Horsfield's hawk-cuckoo (*Cuculus fugax*) led by Keita Tanaka of Japan's RIKEN lab and Rikkyo University. Like the common cuckoo, which has been Stevens's main focus, the Horsfield's hawk-cuckoo kills all its nestmates, whether in egg or chick form. It has a gape that, to humans at least, is a vivid yellow (the host chicks have an orange-yellow gape) and it goes one step further than tarting up its mouthparts. Horsfield's hawk-cuckoo chicks have a yellow patch on the skin on the underside of each wing. This tricks their foster-parents into thinking there are extra mouths to feed; when Tanaka painted a chick's yellow wing patches black, its

parents brought less food. Occasionally foster-parents are so
convinced by a yellow wing patch, even though it's not
mouth-shaped, that they try to stuff food into it. 'The hosts
nest in the dark,' says Stevens. 'They like holes [in the ground]
underneath hanging vegetation so it's probably not necessary
to mimic the shape of the gape. [The patches] are basically a
yellow blob.'

This wing patch is probably unique to the Horsfield's
hawk-cuckoo. But is it a supernormal stimulus? Tanaka
faced a challenge: he needed to work out what a Horsfield's
hawk-cuckoo foster-parent sees. So he and his colleagues
turned to physics kit. First they measured the light bounced
back by a yellow wing patch or gape (their 'reflectance
spectra') using a spectrophotometer – a narrow probe that
analyses which wavelengths of light are present. Tanaka
held the sensor near the chicks' open mouths or by the
patches beneath their wings and recorded the light they
reflected. It was easier said than done. Birdwatching on one
of Japan's most iconic tourist attractions sounds like an
idyllic way to spend a summer, but these birds are 'really,
really difficult to find,' explains Stevens. 'They live in
difficult places to work, high up on mountains, and the
species numbers are low – you might find half a dozen in a
field season.' When Stevens visited Tanaka's main field
site 2,000m (6,500ft) up Mount Fuji for three weeks in
2010, the team found just two Horsfield's hawk-cuckoos,
despite Tanaka's many years of experience.

To track down a rare bird that lives in woods and lays its
eggs inside holes in the ground hidden by vegetation, you
need to learn the places where nests are likely to be based
on the type of plants. 'Then you start looking in those
places,' Stevens explains. 'You find as many hosts' nests as
you can and hope that some of them the cuckoos will find.'
You can also watch out for birds coming and going to and
from nests. 'But basically you just have to try and search for
them all.'

Once the team discovered a nest, Tanaka removed the chicks one by one and made the light measurements while Stevens took photos with a UV-sensitive digital camera. Hawk-cuckoo chicks can weigh up to 90g (3oz) when they leave the nest, whereas chicks of the red-flanked bluetail (*Tarsiger cyanurus*), a common host for the parasite, weigh at most 17g (0.6oz). In total, Tanaka assessed 10 red-flanked bluetail chicks from five broods, and six hawk-cuckoo chicks — five from red-flanked bluetails' nests and one from the nest of a blue-and-white flycatcher (*Cyanoptila cyanomelana*). It took one year to measure the host chicks and three years to find the six hawk-cuckoos.

All agape

These field measurements showed which wavelengths hit the foster-parent birds' eyeballs. But it wasn't the full picture — what did these birds see? To meet the first of the supernormal tests, the cuckoo chick's gape and wing patch would have to be more stimulating to the eyes of the foster-parent than the gape of one of its own chicks. What the bird perceives depends on the sensitivity of its cones and rods (which work well for dim light but don't see colour) to different wavelengths of light. In the same way, a person with full colour vision will spot a number made of red dots hidden in a sea of green dots, but someone who can't distinguish between red and green won't see it, even though the same wavelengths reach their eyes.

The researchers needed a birds'-eye view of the wavelengths their spectrophotometer had recorded. With Stevens advising, Tanaka modelled bird vision mathematically. Knowing which wavelengths reach a bird's retina, which types of cones it has, and how sensitive they are to different colours, the researchers could model what the bird detects and guess what the world (or chick gape, or wing patch) looks like to that bird.

Are you (super)normal?

Stevens assumed that the host birds' sensitivity to different wavelengths was the same as that of their close-ish relative the common starling (*Sturnus vulgaris*), which other vision researchers had measured by shining narrow beams of light onto the visual pigment inside its cone cells and monitoring how much reflects back. 'From doing that and related things, you can calculate how much light each type of photoreceptor absorbs in terms of different wavelengths,' Stevens says. Modern genetic techniques can also indicate cone sensitivity by showing which pigment proteins are inside.

Tanaka's spectrophotometer measurements revealed that the cuckoo chick gape and wing patch are different to the bluetail chick gape. They reflect less longwave light, making them appear extra-yellow to human eyes, but much more UV light, which may help the gape and wing patches stand out in the dark holes where bluetails nest. Putting these measurements into the vision models showed that the gape of the hawk-cuckoo chick is more visible to red-flanked bluetail parents, stimulating their cone cells more strongly than the gape of their own chicks. The same goes for the wing patch, only more so. The parasite chick outflanks its red-flanked rivals.

'[The cuckoo chick gape] is brighter and more vivid than the real yellow chick gape,' says Stevens. 'That's one of the two requirements to be supernormal. First of all you've got to be an exaggerated version of the stimulus – the gape – and second you've got to encourage the hosts to bring more food because of that.'

The wing patch does entice parents to bring more food, as Tanaka showed by painting over it. But to fully confirm supernormal status, the team would need to check how parent birds responded to exaggerated signals, not reduced ones. The role of UV is also not certain. Though there's another clue, this time biological, not physics-based, that something's going on. To display the wing patch, the underside of the cuckoo chick's wings remains feather-free, making it harder for the chick to keep warm. What's more, the birds do a brief but calorie-burning wing shake to show the patch off. Generating bright colours, whether on their wings or inside their mouth, costs yet more energy, this time to create strong pigments from nutrients that could have built another part of the body. So the chances are, whether they're supernormal stimuli or not, the extra-bright gape and wing patch gain these cuckoo chicks at least some advantage. Otherwise they wouldn't go to all that effort.

Tanaka's measurements and the vision analysis assisted by Stevens have proved that the Horsfield's hawk-cuckoo is at least halfway to supernormal. And that's good enough for us. Only by appreciating how different the world looks to a bird that can see UV could the researchers get to the bottom of how these sneaky freeloaders manipulate their hosts. The hawk-cuckoo results indicate that the bright mouths of common cuckoo chicks, although we don't yet have the scientific evidence to know for sure, may also act as supernormal stimuli. The answers may be hidden in plain (bird) sight.

At first glance, cuckoos, ants and bees – the animals whose light powers we've looked at so far – have nothing obvious in common. But on the vision front, they certainly do. These creatures can all see UV, although a more fundamental point is that they all see light that's travelled through air. As do we, unless we're swimming. Then things become less clear – like sound, light underwater doesn't behave the same as in air.

Reflections on water

The sun's blazing, the hay meadow's in bloom and birds are singing by the lake. But you can't hear them and you can't smell the flowers or feel the sun on your face – you're underwater in a world of your own. The diving bug has struck. The only way you can appreciate the sun is to look up. On the calm surface of the water you'll see the whole sky squeezed into a bright circle, surrounded by darkness. It's like gazing into the skylit dome of a dark cathedral. More scientifically, the entire hemisphere of the world above the water is compressed into a cone of light spanning an angle of 97.2° (48.6° to either side of the vertical). The bright circle on the water's surface is known as 'Snell's window' in honour of Willebrord Snell (1580–1626), the Dutch astronomer and mathematician who came up with a law to describe how much light bends – or refracts – when it enters water.

Snell wasn't the first to try to understand this change of path. That honour goes to the Greco-Egyptian scholar

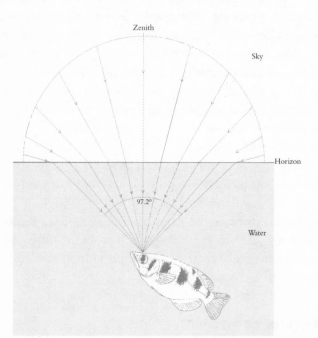

Figure 6.2 **The big picture**. *Light bends, or 'refracts', as it passes from air to water, creating Snell's window, which lets divers and animals underwater see the whole sky squeezed into a bright circle above them.*

Ptolemy. In AD 150 he devised a formula linking the angle that light strikes water to the angle it travels through the liquid. Whenever light moves from one medium to another where it changes speed, it refracts, unless it hits the boundary square on. Light in air zooms at roughly 300 million metres (186,000 miles) per second. Nothing goes faster – one of Einstein's great insights. But in water, light slows to a mere 225 million metres (140,000 miles) per second. If light hits the interface between air and water at an angle, that deceleration bends the light towards a line perpendicular to the water's surface. It travels more steeply downwards, if you like.

To picture why, you can think of the light beam as a line of identical zombies shuffling forwards, each with an identically sized arm on the shoulders of the zombie in front. The rules of this fictional world say that each zombie must – like the peaks and troughs of light waves – stay the same distance apart. All's going well until the zombies, who are groaning and shedding bandages, approach a line on the floor at an angle to it. When the front zombie steps over the line, it must slow down, for reasons of zombie biology too complex for us to go into. The other zombies, unhindered by this biology until they too reach the line, carry on at the same speed. To keep the faster-moving zombie behind at arm's length and avoid a zombie pile-up, the head zombie must rotate its direction away from the original path. As the next zombie crosses the line, it too must slow and rotate, and the column of zombies, one by one, changes direction at the line. If, on the other hand, the zombies march head-on towards the line, the zombies behind the line don't tend to move ahead of the ones who've crossed over, and they shuffle on in the same direction.

Perhaps because he'd never met a zombie, Ptolemy's formula – linking the angle the light strikes the water at to the angle it makes in the liquid – was wrong. Credit for the correct answer should go to tenth-century Persian genius

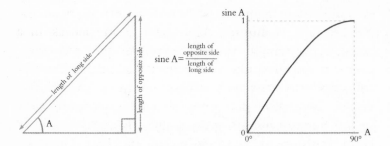

Figure 6.3 **Triangular thinking**. *The sine of an angle is a number that ranges from 0 to 1 and is vital for knowing how much light bends as it passes from air into water.*

Ibn Sahl. But it doesn't. Instead his formula is known as Snell's law, after the man who came up with the same answer hundreds of years later. For light travelling from air into water, Snell's law says that the sine of the angle that the incoming light makes with the vertical is 1.33 times the sine of the angle the light makes with the vertical once in the water. The sine of an angle ranges from 0 for 0° to 1 for 90°, so according to Snell's law, as well as the zombies, light going vertically into water doesn't bend.

So what does this mean for our lake? Complete the sums, and you will find that light striking the surface at 30° to the vertical travels on in the water at 22° to the vertical; if it enters the water at 60°, it continues at 41° to the vertical, and when the light's moving almost horizontal to the surface, it bends to 48.6° in the water. That's why the cone of light underwater extends 48.6° either side of your head yet contains all the light from above.

Bend it like Snell

Refraction is handy if you wear glasses or contact lenses as it bends the incoming light to fall as a focused image on your retina. It causes problems, however, for those living on the edge, where the worlds of water and air collide.

As you'll know if you've tried to fish your trunks from the bottom of a swimming pool, catch a stickleback in a jam-jar on a string or tickle a trout, your quarry is always deeper than you think. Our eyes and brain don't allow for refraction; they assume that light from the shorts or fish didn't change direction, and the shallower angle it now makes through the air is the angle at which it made its whole journey.

Yet archerfish, seven species from the *Toxotes* genus that live in rivers, estuaries and lakes in South-east Asia and Australia, manage to shoot insects sitting on leaves and branches in the air above even though their eyes are below the surface of the water. Creamy-white with brown-black spots or vertical bars on their sides, these fish, which are 10–15cm (4–6in) long, strike with a mouthful of spit fired in a precision-aimed jet. Despite the bending of the light, the animal hits the mark nearly every time, sending the insect plummeting to the water where it can be gobbled up alive. Archerfish are champion spitters. They can fire a drop of water up to 10 times their body length, which is like us spitting 20m (66ft). The *Guinness World Record* for human spitting is, at the time of writing, 33.62m (110.3ft) – an award held by Brian 'Young Gun' Krause of Michigan, US, several times winner of the International Cherry Pit-Spitting Championship.

But the archerfish is top spitter in our book as it's got to be harder to knock a small insect off its perch from underwater than it is to propel a cherry pit forwards in a vaguely straight line. To make matters tougher for the fish, refraction at the surface makes its target appear higher than it is (unless the prey is directly overhead). It's the reverse of one of us trying to nab that fish in a jam-jar, when the animal appears closer than it is. If the light from the insect strikes the surface at 25° to the vertical, it enters the water at an angle of just 10° to the vertical. It doesn't sound a big difference, but if the fish traces back the path of the more steeply travelling light that reaches its eyes, the insect would appear to be higher up. An insect 1m (39in) above the water's surface, for example, would look 35cm (14in) higher than it really is.

Snell–o–vision

So why doesn't an archerfish spit at insects directly overhead, where there's no refraction and the prey is exactly where it appears? Archerfish do that sometimes, but generally prefer not to. Not because they're worried about an insect landing on their head but because an archerfish that wants to spit straight up must point its whole body upwards too. That's a precarious position – if a bird above water (or a rival fish below) spots you, you've got no escape route. Being more horizontal means that the fish can get away faster. Left to their own devices, archerfish will shoot at any angle between 45° and 110°, but their preferred angle is 75° to the horizontal. Spitting at an angle has another benefit too, as it lets the fish reach insects on top of horizontal branches.

Whatever the angle, these fish are incredibly sharp shooters. They almost always bag their target first time; the strike rate of one species is more than 94 per cent. Overshooting, then correcting their aim next time, is not the way of the archerfish. But how are they so good? Are archerfish born with their refraction-busting skill and ability to allow for the gravity that arcs the path of their spit downwards, or do they have to learn? For an answer, let's return to Shelby Temple from the University of Bristol.

Spitting out the solution

As an eight-year-old in Canada in the 1980s, Temple saw a TV documentary about archerfish featuring Britain's top wildlife film-maker David Attenborough. 'He was somewhere in Borneo and I immediately wanted to see an archerfish myself,' Temple recalls. 'It was only when I was 12 that I realised you can buy them in your local pet shop.' Happily, that revelation didn't curb Temple's interest and his quest to ferret out the archerfish's secrets eventually took him from Canada to Australia.

At the University of Queensland in Brisbane, Temple and colleagues discovered that the archerfish *Toxotes chatareus* is such an expert marksman because its retina is adapted to life near the interface between water and air. The lowest part of the retina has up to 50,000 light-detecting photoreceptor cells crammed into a square millimetre, 10 times the density in the rest of the eye. The fish has evolved so that all the light from above the water, which is compressed into Snell's window, lands on this super-sensitive part of its eye. When an archerfish spots a tasty insect, it fixes the image of its prey in this area by keeping its eye as still as possible while rotating its body to fine-tune the destination of its spit. This part of the retina also sees red, green and blue so it's good at detecting insects on leaves. The rest of the retina, meanwhile, ensures that the archerfish is not at risk of attack itself, a big danger for a small fish near the surface of the water, where it's easy to reach from the air and brightly lit for those lurking below. The middle of the retina is great with blue and red light, for detecting birds silhouetted against the blue sky. The area at the top, on the other hand, works best for green and yellow, so the archerfish can spot bright or dark objects, such as a predatory fish, in the murky brown waters beneath.

While in Australia, Temple realised that the only way to find out whether sharp-shooting archerfish are born or made would be to watch them take their first ever spit. If they hit the target first time, their ability must be innate, like those turtle hatchlings in Chapter 5 heading straight to the ocean, a caterpillar making a cocoon, or a bird building a nest. If the archerfish newbies need a few goes, they must learn by trial and error.

Fishing for results

But how to get hold of freshly born archerfish? Temple's solution was to head out from his lab, roughly halfway up Australia's east coast in south-east Queensland, to the Laura River further north. There he and a local fish-breeder captured *Toxotes chatareus*. Temple shipped the fish home and waited for them to lay eggs. He ended up with 35 newborns but, having promised the lion's share to his fishy friend, he was left with just nine, to which he gave a tank each. That's not a lot of fish to play with, so the pressure was on to get results. Archerfish can spit once they're around a week old and 2cm (0.8in) long. To stop the youngsters spitting prematurely, Temple placed a layer of bubblewrap on the water in each tank. When it was time, with video camera on standby he whipped back the bubblewrap of one tank and put a fruit fly on a leaf 10cm (4in) above the water. It was a tense moment. Would the archerfish spit on or off target when it had never spat before?

The results were underwhelming. No majestic jet of fluid, just a small dribble. 'Its spit went barely 1cm [less than half an inch] above the surface,' Temple says. 'It was the cutest thing ever though.' So did that mean archerfish start life bad at spitting, then improve? Not so fast – Temple had forgotten that most fish only grow as big as the environment they're in allows. As the tank was small, the fish was still a tiddler by archerfish standards, making the finding null and void. Temple needed a full-size archerfish to know for sure. But there was a snag. Of the nine fish he'd started with, seven others had either leapt on top of the bubblewrap or decided to jump – not spit – to catch their prey. Only one remained. It had a lot resting on it. Temple didn't give it a name, but let's call it Archie.

Temple moved Archie to a larger tank and waited another couple of months. When the big day finally came, he removed the bubblewrap and ... Archie hit the target first time. This final fish became a celebrity, appearing on two Australian TV shows. So archerfish can spit accurately, correcting for refraction, their very first time? Sadly, we can't say for definite. In science, you don't draw conclusions from just one example, and Temple had no fish left and no grant money either. 'It's open to debate, but my hunch still is that archerfish have evolved to spit accurately,' he says. 'It's not something they learn.' We still don't know exactly how the fish account for refraction. It's all very fishy.

Into the red

Most other aquatic animals live deeper down than archerfish and don't need to deal with the refraction as light moves from air to water. They have other problems.

In a water-filled trough, a small *Japetella heathi* octopus sits in the unlit gloom of a lab on board a ship floating off the west coast of South America. Around 5cm (2in) long, the creature looks like a Mr Potato Head toy, with two large eyes seemingly stuck onto a lumpen oval body and legs forming his unruly hair. Suddenly a blue light flashes and the octopus turns red, like a blushing teenager at a disco.

It's no surprise that this octopus can change colour; biologists have known that for years. The octopus is a master of disguise; it has a lot of legs to stand on when it comes to altering its appearance. An angry octopus, as deep-sea vision expert Sönke Johnsen of Duke University, US, explains, goes into a white rage, turning pale all over apart from dark circles round its eyes, like Hector the ghost. 'It's a pretty forbidding look,' Johnsen says. An attacking octopus changes from brown to black just before it strikes. A male cuttlefish – a close relative of the octopus – that's wooing a female will pull out all the stops and display rippling waves of colour. More impressive still, this chameleon of the seas may show a decidedly one-sided approach to male beauty, keeping the half of his body that faces away from his potential mate camouflaged; mating displays are like walking a tightrope between luring in a partner and attracting an animal that will eat you. So it may not be a surprise that cuttlefish and other cephalopods (from the Greek for 'head-feet') such as squid and octopus are the most intelligent invertebrates by far. When the competition includes slugs, snails and ticks, that might not astound you. But if you place a broken shape behind a cuttlefish, as Sarah Zylinski, a former colleague of Johnsen who is now at the University of Leeds, UK, has done, the animal will camouflage itself to the pattern by filling in the gaps.

'Working with octopus is fun,' says Johnsen. 'You can see what an octopus is thinking because it appears on their skin. We blush, we make faces and so on. They put up almost an entire movie on their bodies, which tells you what's going on in their heads. They're fascinating.' The animals change colour by stretching or contracting the muscles of their pigment sacs, or chromatophores. When contracted, the chromatophores form pinpricks of colour, but stretched out they make much larger spots. It's like having freckles that you can grow and blotch together at will.

The excellent communication skills of the octopus aren't just good for the animal; they make studying its vision easier too. 'It's been very difficult for us as biologists to get into the minds of other animals,' Johnsen says. 'Octopus … will put patterns on their body that will let you know what they're seeing.' Like we said – clever. Although nobody knows why octopodes (the official plural as the name comes from the Greek for 'eight' and 'foot') are so brainy. 'The smart animals are pretty much always the social ones,' explains Johnsen. 'Intelligence has evolved to allow animals to deal with a complicated social situation where you have to remember individuals, how they treated you and who's important and who's not. All the things we do on a daily basis.' Yet octopus are entirely asocial, and squid live in schools but don't appear to have a complicated social hierarchy. This impressive intelligence in a solitary animal intrigues Johnsen.

Reddy steady go

If the octopus is so clever, it must have a reason for 'blushing' when the light turns blue. The answer, Johnsen believes, lies in what happens to light underwater. If you're a diver, you'll know about Snell's window and the problems of the archerfish at first hand. And if your diving buddy is pale, you'll have noticed their face turn green and their lips blacken as you sink deeper. Just as the world of sound changes underwater (see Chapter 4), so too does the world of light. Water is denser than air and it absorbs and scatters light more. The colours change because the liquid absorbs more of light's longer wavelengths, which have the right energies to give extra oomph to the water molecules' vibrations and rotations. The energy of the lightwave converts to heat energy in the molecule. So the longer wavelengths – red, orange and yellow – travel less far than the shorter-wavelength violet, blue and green light. That's why as you dive deeper below the surface, pink- or yellow-tinged faces become green, while red lips, or blood, look black. It's also why plants such as single-celled algae and phytoplankton, which need red light to photosynthesise, can only survive in the upper 200m (660ft) of the ocean.

If you're a fish living in open water where light is relatively plentiful, there's nowhere to hide. Many fish get round this potential danger by being transparent so they're invisible to predators looking up from below, who would otherwise see a dark silhouette against the light filtering down from above. Other fish at these depths take a three-pronged strategy to cover predators from every angle: they have photophores – light sources – on their undersides so they don't create a shadow for predators below; a dark back to stop them standing out against the murky waters beneath for creatures looking down from above; and silver flanks to reflect light and render them invisible to animals to the side.

Travelling light

As it travels deeper, even blue light becomes so feeble that by about 600m (1,970ft) below the surface, divers find it hard to see. Below 850m (2,800ft) they can't see at all even if the water's clear and it's a sunny day. Under the same conditions, no animal can see below 1,000m (3,300ft). Once the sea is largely dark, there's a clear snag with transparency: it can make you more visible. For there are creatures down here, like the flashlight fish, the lanternfish and the dragonfish, that have forged their own solutions to the darkness. And larger species of at least one – the dragonfish – feed on *Japetella heathi* octopus. These deepwater fish have developed flashlights, bioluminescent organs near their eyes that oxidise the pigment luciferin in a chemical reaction that typically emits blue–green light.

A transparent animal's body tissues, even though they're invisible when there's lots of light around because they let almost all light pass through, have subtly different refractive indices. The refractive index of a material is equal to the speed of light in a vacuum (which is close enough for us to its speed in air) divided by the speed the light travels in that material. Water has a refractive index of 1.33, that figure we used in Snell's law in the archerfish section to work out the angle light bends when it moves from air to water. The index also tells us that the light slows to 1/1.33, or 75 per cent, of its speed in air. Crucially for transparent sea creatures, when light hits a boundary between two tissues with a different refractive index, a small fraction of it reflects back.

'Animals using flashlights in the dark can pick up transparent animals,' says Johnsen. 'They can see them because a little light gets reflected back, more than gets reflected from the water itself.' It's like pointing a torch beam at a window at night – the glass is clear in the daytime but in the dark it's given away by the reflection from the flashlight.

Higher in the ocean, there's too much light around for the small amount reflected back by transparent animals to be noticeable; it's like hunting for a reflected torch beam on a bright summer's day. But the only way for a sea creature to be invisible to a flashlight in the ocean depths is to absorb everything from the flashlight that hits it, rather than letting almost all of it travel through. Deep down, it's best to be red or black, not transparent. 'Most flashlights are blue,' says Johnsen. 'So if you're red you are absorbing all the light that hits you.' A red animal appears red because it only reflects red wavelengths, absorbing all other colours. Black animals absorb all the light of any colour. So a red or black animal in the deep sea won't reflect searchlight beams back to the fish they belong to. The fish with the 'torch' will think its beam has petered out in the gloom and there's nothing there; it won't suspect it's missed a body.

Stray higher, where there's more light, as a red or black sea-creature and your dark shape will form a silhouette that's easy to spot from below. To complicate matters, the depth that daylight reaches depends on many things: the time of day, the amount of cloud, how murky the water is, how recently there's been a storm, and more. So the attributes that will make you invisible alter as you move up and down in the ocean and as the ocean itself changes. This set Johnsen wondering. Having seen the way many cephalopods camouflage themselves to their surroundings, he was curious whether they'd take this one step further and alter their appearance in response to a change in the light. Are octopus really that advanced?

Trawling for results

To find out, Johnsen and his team went to sea. One September night in 2010 on a cruise above the Peru–Chile trench, Johnsen's colleague Zylinski trawled several young *Japetella heathi* up from a depth of 100–500m (330–1,650ft). In these octopus, the main body – the mantle – is around 8cm (3in) long. Changing light conditions might be of particular interest to *Japetella heathi*, as when young this octopus tends to live in the top 400–700m (1,300–2,300ft) of the ocean during the day, while when it's older it heads deeper than 800m (2,600ft). Zylinski installed her catch safely in the ship's lab. Then she dimmed the lights and picked out the blue wavelengths (roughly 450 nanometres) from a white light-emitting diode (LED, as found in some low-energy light bulbs) with a filter. She shone her artificial bioluminescent searchlight onto the octopus and shot a video of its reaction.

Sure enough, within about a second of introducing the blue beam, Zylinski saw red. Not because her experiment had failed but because the octopus, until then merrily transparent, changed colour, as we described at the start. The octopus did not respond to a red beam; they showed a clear preference for blue light. Zylinski found the same result for a few juveniles of a medium-sized cephalopod known as the common clubhook squid (*Onychoteuthis banksii*). The squid too turned red when they experienced the blue beam, and ignored the red light. Johnsen realised that this could be a camouflage trick.

Digging deep

That said, an octopus or squid in a lab is likely to act differently to when it's at liberty in the sea. 'These experiments would be extremely difficult to do at depth,' says Johnsen. 'Even if you took a sub down there, your arrival with lights and noise would totally mess up any normal behaviour.' According to Johnsen, this is a problem with all deep-sea biology. 'We don't know for certain what animals do down there,' he says. 'Especially for animals that can move around and move relatively quickly. The moment we've arrived we've disturbed their behaviour.' To solve this challenge, biologists want to put infrared cameras on tripods on the seafloor, though they'd be in for a long wait. 'The deep sea is relatively empty so you can film an awful lot of nothing,' Johnsen says.

To replicate the light levels of a fish flashlight and the deep sea more closely, the team dimmed the LED beam, and the lab lights, some more. The octopus behaved the same but it was too dark to shoot video. 'We'd love to know if this is what's going on at depth,' Johnsen says. 'It would be more accurate to have a very small beam of light rather than a whole flashlight [the LED]. It's hard to separate that from all the lights going on or off. The octopus must ask themselves, is everything brighter because a cloud just moved away from the sun or because a small pencil of bioluminescent light is heading their way?'

Still curious, as all good scientists are, on a cruise in July 2011 in the Gulf of California, Zylinski studied four *Japetella heathi* octopus taken from a depth of around 1,000m (3,300ft) during the daytime. When Zylinski pressed a blunt needle onto one of their arms to stress them, the octopus turned red, reflecting back around half as much blue light as when they were transparent. Although the red octopus reflected back more red light than blue, it was still less than the amount of red light they reflected while transparent. Across all wavelengths, an area just forward from the gut on a red octopus reflected less than one-fifth of the light that hit it, Zylinski confirmed, and between one-tenth and one-twentieth of the blue-green light.

It looks as if this octopus is smart enough to turn red and non-reflective when it strays into the searchlight beam of a bioluminescent fish. For once, a blush serves a useful purpose. Unless your enemy is a dragonfish. This is the type of fish you'd expect to find in a glass case in a Victorian museum, with an open mouth that terrifies, its upper jaw angling sharply upwards to create a yawning cavern lined with sharp teeth. Sneakily, dragonfish use both blue and red light in their searchlights. Although the three species that do this are probably too small to eat octopus. 'In the deep sea it doesn't really matter if you're red or black unless someone turns up with a flashlight of a different colour,' says Johnsen. But that's another story ...

Squids in

Talking of stories, how's this for a tale of deepwater sealife? 'Before my eyes was a horrible monster worthy to figure in the legends of the marvellous. It was an immense cuttlefish, being eight yards long. It swam crossways in the direction of the *Nautilus* with great speed, watching us with its enormous staring green eyes. Its eight arms, or rather feet, fixed to its head, that have given the name of cephalopod to these animals, were twice as long as its body, and were twisted like the furies' hair.' These are the words of fictional French marine biologist Professor Pierre Aronnax in Jules Verne's *20,000 Leagues Under the Sea* (1870). Spoiler alert: skip to the next paragraph if you haven't got round to reading this classic in the nearly 150 years since its publication. Hired by the government to find a ship-destroying sea creature, Aronnax discovers the *Nautilus* submarine belonging to Captain Nemo, who promptly kidnaps him and holds him prisoner on board the sub. No prizes for guessing what the sea 'monster' turns out to be. Later a pack of seven cephalopods attacks the *Nautilus*. Exactly which type of cephalopod Aronnax is talking about is debatable. In the English translation they're giant squid or cuttlefish while in Verne's original French they're octopus (neither giant squid, as far as we know, nor octopus hunt in packs but it's a good story even if the biology isn't spot on). Whichever type they are, many of the cephalopods die when Canadian whaler Ned Land plunges his harpoon into their massive eyes.

Reading about this submarine–squid showdown as a child inspired Johnsen to study marine biology, though not until after he had gained his joint degree in art and maths, for which he claims to have spent most of his energy on the art part. 'The visual arts are important to me,' he says. 'When I went into biology I always knew that I would be working on things that involved light and colour.' Sure enough, these days Johnsen specialises in animal sight.

'Like most people squid are always in the back of my mind as the great monster,' Johnsen says. 'It's hard not to be interested in squid eyes – they're so much bigger than every other eye out there, even for animals that are larger.'

As Verne described, the giant squid (*Architeuthis* species) is a formidable enemy. Apart from its eight long arms, this animal has two even longer tentacles that trail out from its body. All 10 appendages are covered in suckers, and it has a pointed beak over its mouth for killing and tearing up prey. The squid looks like a cross between a creamy-white chunky jellyfish and a floppy octopus. On a massive scale. The giant squid lives up to its name – it can be 13m (43ft) long (about 5m (16ft) to the tip of its arms, with the tentacles adding the rest), making it the second-largest invertebrate. Females, which are larger than the males, can weigh nearly 300kg (660lb), more than a male lion or big pterosaur.

The giant squid and the roughly 1-metre-longer colossal squid (*Mesonychoteuthis hamiltoni*), which is more pink in colour and the largest invertebrate of them all, have eyes as big as dinner plates. That's nearly three times the diameter of any other animal's eye. Both the giant squid and the colossal squid (squid namers are not known for their imaginations) follow the trend for deep-sea gigantism where, for reasons known only to themselves, animal species that live at depth grow larger than their shallower-dwelling cousins. The next biggest eye in the animal world is just 9cm (3.5in) across, about the size of the giant squid's pupil. This runner-up position belongs jointly to the blue whale (*Balaenoptera musculus*), at 30m (100ft) long the largest animal of all, and the roughly 3m (10ft) swordfish (*Xiphias gladius*), which lives at depths of up to 550m (1,800ft). Our eyes are a titchy 2.4cm (0.9in) across; the largest eyes on land are those of the ostrich, which has 5cm (2in) orbs.

Large eyes sound great, but they're biologically expensive. 'You need a tremendous amount of blood flow to keep cone cells and rod cells working,' Johnsen explains. 'And having a large eye means you need a big piece of brain devoted to processing the image. Any time you have a big eye you have to wonder what this big eye is for, because it's paying such a large price.' Johnsen, together with Eric Warrant and Dan-Eric Nilsson, both at Lund University in Sweden, had a hunch that big squid get a payback for their jeepers-sized peepers.

Big eyes let in more light. They can also contain more photoreceptors, so they are more sensitive to that light and pick out fine detail better. But underwater there are diminishing returns once your eye is bigger than a certain size, because the water absorbs and scatters light so much. What makes these squid so special? The answer lies in tactics that once saved the life of a fighter pilot who went on to become an astronaut.

What big eyes you have, Grandma

Until Johnsen, Warrant and Nilsson completed their research in 2011, the giant eyes of the giant and colossal squid were a mystery. As in many ways are the lives of these massive invertebrates. The animals live so deep in the ocean, probably between 500 and 1000m (1,650 and 3,300ft), that we hardly ever see them in their natural habitat. We know that the colossal squid is a chunkier, as well as longer, chap than the giant squid, with the females weighing up to 500kg (1,100lb). The colossal version has hooks, as well as suckers, on its tentacles and hangs out in the Southern Ocean, in a band stretching around Antarctica and as far north as the southern tip of South Africa. The giant squid is found everywhere around the world but has a particular fondness for the waters off New Zealand. We say 'found', but that discovery is rare – occasionally we come across them stranded on beaches, or drag them up with fishing nets or lines, or see their remains Jonah-like inside the stomachs of sperm whales (*Physeter macrocephalus*).

If Johnsen thought studying octopus was a challenge, investigating squid was even harder. 'Even if you're a practising marine biologist like me, who goes to sea regularly, I will probably never see one alive, which makes doing the biology very difficult,' he says. 'At least with these other animals you can bring them on a ship and experiment with them. Even if you ever got [a giant or colossal squid] on a ship it would be too large to do any experiments with.' Whales are easy to work with in comparison, Johnsen says – there are more of them, they're easier to find as they have to come up to the surface to breathe, and you can tag and track them. 'With giant squid there've been [only] one or two cases where people have got any video footage at all, let alone have been able to study them.'

Without easy access to live giant or colossal squid, the researchers turned to photos and a frozen eyeball. In 2012 Warrant and Nilsson examined a photograph of the eye of a giant squid found near Hawaii in 1981. They brought the photo, along with the original measurements from the dead animal, to Johnsen. This squid had an eyeball at least 27cm (10.5in) across and a pupil, at 9cm (3.5in), about the width of a large hand. Johnsen was also fortunate enough to obtain the defrosted eye of the largest colossal squid ever discovered, which was picked up by a New Zealand fishing boat. This eye was similarly sized, between 27 and 28cm (10.5 and 11in) – about 5cm (2in) bigger than a football.

'This was a classic case of forensic biology because we didn't even have a living animal,' Johnsen says. 'We had the size of the eye. We were able to take measurements of how big the lens was, how big the pupil was … We developed a mathematical model for how vision works underwater and used that to explore what advantage a big eye might have.'

As we saw with the octopus, water absorbs light much more than air does, particularly at longer wavelengths. In addition, the water molecules cause scattering, absorbing and re-emitting blue light the most, sending it out in all directions and deflecting it from its original path. At shallower levels, zooplankton – floating animals just visible to the naked eye – reflect all wavelengths, making them appear white. Phytoplankton (floating plants often too small for us to see) and dissolved matter or other tiny particles reflect back light of all colours too, acting like raindrops in a mist that blur vision. Trying to see through water is like peering through fog. Researchers accustomed to seeing through air must employ their imaginations and science to understand what vision is like for an animal in water.

'When you're on land and your friend walks away, your friend will likely vanish to a point or behind a building well before they fade into the mist,' Johnsen explains. 'In the ocean it's the opposite. Animals almost always will fade away before they get smaller and shrink to a dot.' This is one of the disturbing things about doing underwater research, Johnsen says: something can be the size of a 747 Jumbo Jet and only 3m (10ft) away, yet it's impossible to see. 'That's true underwater all the time,' he adds. 'Even if the water seems quite clear, you're never going to see more than about 100m [330ft]. And it doesn't matter how much light you bring with you. In the submersibles we have intensely powerful lamps but eventually they just send more and more light back to you [because of back-scattering]. Under clear conditions, we normally only see 20–40m [65–130ft] at most.'

Each underwater creature effectively lives in a bubble: it sees a fixed distance before objects become indistinguishable from the murk. How far it needs to see depends on its size. If you're a small fish that only has to find food smaller than you, avoid your slightly larger predators and seek another small fish to mate with, then seeing about 10m (33ft) is plenty. 'You don't need to see another small fish 100m [330ft] away, it has no impact on your life,' Johnsen says. For a copepod, a small crustacean, this distance of interest is just a metre (3ft). That's why small sea dwellers don't bother having massive eyes that let in more light and see further underwater. 'The ability to see a long way only makes sense if you care about things that are still pretty big [at that distance],' Johnsen explains.

A whale of a time

With their record-breaking eyes, the giant and colossal squid must be looking to spot a big animal looming in the dark a long way off. But which one? Captain Nemo's submarine aside, almost the only thing these deep-sea-dwelling squid have to fear is the sperm whale, which likes squid suppers. Squid of all sizes, not just the giant and the colossal, make up four-fifths of the sperm whale's diet; it tops up on fish. Some sperm whales have scars on their skin from squid suckers, and the ambergris that collects in their stomachs, which we used to put in perfume, could be a reaction to damage from pointed squid beaks. Apart from squid beaks and suckers, in past times the sperm whale was mainly under threat from candle- and lamp-makers after its spermaceti, the liquid wax in its huge, flat-topped head. These days the species is protected. Mostly.

Sperm whales dive to more than 2km (1.2 miles) deep to hunt fish and squid, making them the second-deepest-diving mammal after the Cuvier's beaked whale. A male sperm whale is around 16m (52ft) long – a fraction longer than the squid – and hundreds of times heavier at 40,000kg

(40 tonnes). That's heavy bones for you; the squid, as an invertebrate, doesn't have any. The sperm whale's body is also much thicker, Johnsen adds. 'Most of the length of a giant squid is stringy arms and tentacles; sperm whales are a mountain of meat.' The sperm whale may be a tad longer and much heavier, but were it to have an eyeballing contest with a giant or colossal squid, the squid would win fins-down. The sperm whale's eye is a miserable 5.5cm (2.2in) wide. We don't know how far sperm whales can see or even if they can see forwards – their huge head may get in the way.

Although this whale may not be able to see far in the murky ocean depths, it has another weapon up its sleeve. The sperm whale gives out, at 230 decibels (relative to 1 micropascal at 1 metre, the standard reference for sound in water), some of the loudest sound waves of any animal, and listens – like the bat on land – to the echoes bouncing back to track down prey. The whales have to 'shout' so loudly because the squid's soft bodies don't reflect sound well. Even so, sperm whales can detect the 25cm (10in) *Loligo* squid up to 325m (1,100ft) away and the muscular 1.5m (5ft) Humboldt squid at 1,000m (3,300ft). The giant and colossal squid are more flaccid so they're likely to be harder to find with this sonar. But the chances are the sperm whale can sense them from more than 100m (330ft), making the squid's 'bubble' of interest at least that big. The squid can't hear the sperm whale's sonar clicks – their main components are at a frequency of 15,000Hz, above the squid's detection range. Since the squid don't have their own sonar system they must find another way to avoid becoming dish of the day. All they have to rely on are their eyes, although seeing more than 100m (330ft) underwater is an almost impossible feat.

Talking of winning against the odds, one night US navy pilot Jim Lovell was flying home over the sea to his aircraft carrier. All was well until his navigation system failed and he had no way of finding the ship. With the presence of mind that would later serve him well as an astronaut on Apollo 13, Lovell remembered his marine biology. He turned off his plane's cabin lights, then peered into the darkness. In the distance he could see a faint blue-green glow. Disturbed by the wake of the ship, small bioluminescent organisms such as bacteria and dinoflagellates were firing out pulses of light that guided him back to safety. For the same reason, you can sometimes see a dolphin highlighted by twinkling 'dust' as it glides through the water. This is also the tactic that the giant and colossal squid use to protect themselves from distant sperm whales, Johnsen, Warrant and Nilsson's research shows. Almost no sunlight penetrates this deep, so the squid's chances of seeing a sperm whale silhouetted against light from above are slim. But a moving sperm whale creates tiny dots of bioluminescent light like glitter in a snow dome. There's just one snag: the dots are small and far away. For the squid, spotting these pinpricks of light is equivalent to seeing a shape in the static of an old-style TV when your living room is large and full of fog. Could the massive eye of a giant or colossal squid be sharp enough to detect these whale-induced sparkles?

Light rain

It's a phenomenal task. When you're trying to see a faint glow in the dark, it's all about the statistics. 'You're collecting light from the object and you're collecting light from the background,' says Johnsen. 'Ultimately you have two numbers. You need to know that those two are different, to reliably say there's something there.' On land distinguishing something from its background is easy unless it's foggy or there's not much light. Then it's like trying to read a book in the dark. Although you can see the book, you probably can't see enough contrast between the white of the page and the black of the text to pick out words. 'As it gets darker it becomes more and more obvious that light isn't all arriving like a smooth river, it's arriving like raindrops, what we call photons,' Johnsen says.

As we mentioned in Chapter 1, photons are the smallest amount of light possible; under certain circumstances light no longer behaves like a wave but like a series of tiny packets, or particles, in what's known as wave–particle duality. Scientists argued whether light was a particle or a wave for more than a hundred years (Newton was on the particle side of the fence) before realising that it's both – an excellent compromise.

'The light is arriving in the eye bip, bip, bip [like the sound of a heartbeat monitor in a hospital drama], little bips one after the other, and it's random,' Johnsen says. 'Imagine you drew a circle on the sidewalk in chalk and then it started to rain and it rained a little bit more in that circle than around the circle. Suppose only four drops had fallen, would you be able to tell that circle is there? Probably not.' Even after a hundred drops, more of them landing inside the circle than out, you probably wouldn't be able to see the shape. 'But now imagine thousands of drops fall and the area of the circle is much wetter than the area around it, then you can see the circle really well,' the researcher says. 'That's the trick. The animals need to be able to collect enough light so they can get around the statistical problem of knowing if something's there or not.'

The way to get enough light is to have a big eye. Squid have simple, camera-like eyes like ours, with a single lens that focuses the light that enters the pupil onto the photoreceptor layers of the retina. 'You need a big eye, you need to let a lot of light in and you need lots of photoreceptors, many visual cells recording a lot of data to get the ability to see things that are just a little bit lighter or darker than the background,' says Johnsen. The researchers' models showed that giant or colossal squid can detect tiny differences in contrast well enough to spot the blue-green glow from bioluminescent plankton 120m (390ft) away. That enables the squid to monitor a whopping 7 million cubic metres (250 million cubic feet) – around 2,800 Olympic-sized swimming pools' worth – of water at once for sperm whales. The squid's eyes are better at sensing the blurry low-level glow from a collection of these light dots than spotting them as individual points, Johnsen discovered. It's like seeing the blur of the Milky Way rather than resolving it into individual stars, or detecting the glow from a string of fairy lights 100m (330ft) away instead of each single bulb. The big eyes of the squid are unlikely to help them spot other squid of the same species or their prey – typically fish and smaller squid – as these animals aren't large enough to make a glow over a sufficiently big area.

The only advantage to an eye as huge as that of the giant or colossal squid is its ability to detect a massive, faintly luminous object a long way away below 500m (1,650ft) or so, where there's not much light, according to Johnsen's calculations. It looks as if the mystery of the giant squid eye is solved: it's extra-big to spot the glow of distant whales swimming through bioluminescent organisms. Although there's still a chance that the squid's large eye merely detects a dark shadow from up to 100m (330ft) away. Or, more likely, it's doing both. It's hard to prove. In shallow water, beyond a diameter of about 10cm (4in), having a bigger eye doesn't help you see much further, which goes some way to explain the gap between squid eyes and their nearest competitors.

Time to squidaddle

So the squid can see the sperm whale from roughly the same distance that the whale can track the squid with its sonar beam. That puts the animals on a more equal footing, but the squid still has to get away. Given that the sperm whale's top speed is an impressive 32kph (20mph), it's unlikely that the squid will outswim it. But the squid can exploit two things to escape: a flaw in the whale's sonar system and the optical properties of seawater. The whale's sonar is directional. A squid that spots the glow from a distant whale and moves out of its detection line may be able to evade the sonar beam and quietly fade into the 'fog' of the light-scattering water. 'If you were chasing a car on the highway, that car could be a mile away and you'd still be able to see it,' Johnsen says. 'But underwater it's a different world because once something's more than 100m [330ft] away you're not going to be able to see it.' It's like a game of hide and seek in the dark in thick fog, with the whale able to 'see' only where it shines a narrow torch beam, and the squid able to see through the mist in all directions. 'If the giant squid can swim off in a different direction it may never be encountered again,' Johnsen says.

The only other animals with eyes as big as the colossal and giant squid's are now extinct. Ichthyosaurs – large marine reptiles or 'fish lizards' that lived around 250–290 million years ago – were up to 16m (52ft) long. Their eyes may have been 35cm (14in) across, equivalent, if we stick with the crockery analogy, to the charger found underneath a standard-sized dinner plate in a fancy restaurant. Again, it's a case of 'large animals trying to see other large animals through the water'. Whales didn't yet exist, so the ichthyosaurs were probably looking out for each other or for viciously toothed pliosaurs.

Much about the massive squid is still a mystery. But what would Johnsen like to find out about their vision? 'That's before we've seen the thing in the wild and how it behaves and what it cares about,' he says. 'Barring that, we would like to know more about its retina. We know about the shape of the eye, the pupil, the lens, we know the optics of the eye, we know how it works as a camera.' Discoveries about the retina could reveal if the squid has colour vision, how sharp its vision is, how it processes the information in its brain, what it can see and what it can't see. 'We know the simplest parts but we don't know the parts that would let us know in detail what its perceptual world is like,' Johnsen says. 'For giant squid we know almost nothing.'

A light round-up

In this chapter, we've seen animals that use the physical properties of light in a myriad of ways: ants and bees exploiting polarisation to find their way, cuckoo chicks with UV wing flashes to gain more food, archerfish that correct for refraction so they can knock insects into the water, octopus changing colour to hide from searchlight-wielding fish, and massive-eyed squid that escape sperm whales by spotting the bioluminescent glow from distant plankton disturbed as the whale swims by.

These animals showed us that light is a wave made up of oscillating electric and magnetic fields at right angles to each other. Ants and bees revealed that the sky polarises these fields and how scattering makes that sky blue. Horsfield's hawk-cuckoo chicks and their yellow and UV-reflecting wing patches showcased that light has a wavelength and frequency just as sound and other waves do, while the archerfish demonstrated how light bends – or refracts – as it enters water and changes speed. Octopus helped us with more on colour, reflection and refraction. Finally, the colossal squid brought the giant news that light can act both as a wave and as a particle.

It's best to go out in style, so what's more appropriate than the largest invertebrate of them all finishing our whistlestop tour of animals that exploit physics in their daily lives? There's nothing like a big ending. Although we're not quite finished yet. Read on to discover whether the animals know what they're doing, and the darker side of animals and physics.

Life, the Universe and Everything

A life of physics

Our world is full of animals using the principles of physics to survive. Archerfish shoot down flies sitting on leaves above the water; eels stun their prey with electric currents; and red-sided garter snakes huddle together in their thousands to conserve heat (and steal it, in the case of she-males). Even everyday animals have physics nous. The domestic moggy uses surface tension to lift liquids up with its tongue, while dogs shake themselves dry by following the rules of simple harmonic motion. Filling the book was easy, as there were so many options. But that also made it tough, as there were lots of creatures we didn't have room for (so apologies to the 'burglar-alarm' jellyfish, the polarisation-cancelling herring and the electric-field-sensing duck-billed platypus).

Physics isn't, however, only handy for understanding how animals communicate, defend themselves, move, eat or drink. It's also given us equipment for studying animals in action. Ammeters for measuring the charge on bees. Wind tunnels for studying how they fly. Infrared cameras for mapping the heat from a California ground squirrel's tail or the blood droplet on a female mosquito's bottom. The surprising tech star of this book is the high-speed video camera. By capturing fast movements at hundreds or thousands of frames per second, and watching the footage back in slow-mo, scientists have learned everything from how mosquitoes survive raindrop collisions to the way pondskaters scoot across the surface of water.

Along the way, *Furry Logic* has showcased many great physicists. But to us there was no bigger genius than Isaac Newton. Forget Albert Einstein and his mind-bending musings about the speed of light. When it comes to the 'classical' physics governing animal life, Newton's our number one. He has cropped up in almost every chapter with his development of the laws of motion, his studies of liquids and his calculations of the speed of sound (even if he was 20 per cent out). Newton was an oddball – he spent years trying (and failing) to turn base metals into gold and held controversial religious views – but his scientific vision was incredible, even if he did risk his own sight by sticking things down the side of his eyeballs to see how they altered his view of the world.

Inspired by nature

Yet physics doesn't have a monopoly on wisdom. Ideas have often flowed in the opposite direction, with the behaviour of animals triggering advances in physics. Ancient Greeks pondered forces while watching oxen pull carts. Eighteenth-century scientists grappling with the electric eel's shocks led us to understand electricity. Horses turning a bore proved that heat is a form of energy, while Germany's Otto Lilienthal (1848–96) made the world's first glider after studying storks.

Animals have inspired new technologies too. Sometimes this has simply been researchers building artificial animals to learn about the real ones: we saw the Robostrider pond skater, the robotic hawkmoth and the stuffed California ground squirrel with industrial heaters up its tail. More significantly, scientists and engineers have developed technology that mimics animals, including a vibrational hearing aid inspired by African elephants, and adhesive Geckskin pads that copy gecko toes. Then there was Newton's suspicion that peacocks get their beautiful colours not from pigments but from the way light bounces off their feathers. That insight – since confirmed – has led to an entire new science of 'structural colour', which is creating everything from hard-to-forge banknotes to smart windows that respond to the environment, and even a dipstick that tests how much alcohol is in a drink.

This two-way flow of information shouldn't be a surprise. It is mere convention to talk about biology and physics as if they're unrelated; they're just labels we give to different ways of looking at nature. Convenient, but not necessarily helpful. Dividing physicists and biologists – making them go to separate classes and learn different subjects – stifles progress. Each camp ends up speaking a different language: to a physicist, a nucleus is a collection of particles at the heart of an atom; to a biologist, it's a structure at the heart of a cell that contains genes.

Many physicists are guilty of believing that everything reduces to physics. What is an animal, they will say, other than a collection of atoms and molecules made of electrons, neutrons and protons, themselves composed of quarks and gluons? That's true, but it only gets you so far. Though we use the movement of air molecules to explain *how* peacocks create infrasound, we won't know *why* they make those noises unless we study their mating habits. The world's a complicated place that can't always be boiled down to physics; and that's without even mentioning animal genetics, neuroscience or physiology.

And physics isn't entirely benign. Mankind's technology, much of it based on physics, has damaged many animal habitats. Coastal power lines and steel in beach-side hotels could disturb the magnetic fields that loggerhead turtles use to navigate home. Mining for coltan, the ore that provides tantalum for capacitors in our mobile phones, laptops and other electronic gizmos, in the Democratic Republic of the Congo destroys wildlife habitat and promotes hunting of eastern lowland gorillas. By enabling some humans to flourish, technology has boosted our population so high that we are encroaching on wild landscapes with our farming and cities. And the Industrial Revolution kick-started our burning of fossil fuels that is now changing the climate.

Used wisely, though, technology can be a force for good by improving health care, providing safe water and giving us clean energy that doesn't emit carbon dioxide or pollutants. And it can help us to conserve the animals we've landed in trouble. We mentioned, for example, using elephant seismic calls to lure marauding males away from conflict with humans, while knowing that turtle hatchlings imprint on their beach's magnetic field could mean we can reintroduce turtles to coasts where they've become extinct.

Knowing me, knowing you

Before we end, there's one elephant in the room (or should that be in the book?) that we need to address. Do animals actually 'know' physics? We've talked throughout as if animals use physics consciously (partly because it's easier to tell a story by anthropomorphising). But can animals reason about abstract notions such as mass, gravity, forces and material strength? It's a question that Daniel Povinelli, an anthropologist at the University of Louisiana, US, tackled in his 2000 book *Folk Physics for Apes: the Chimpanzee's Theory of How the World Works*. Out in the wild, these animals build and use tools: stones and anvils to crack nuts, and sticks to pull termites from their mounds. In the lab, researchers have seen them stack boxes to access bananas hanging out of reach. Similarly, zoologist Alex Kacelnik at the University of Oxford, UK, once showed that New Caledonian crows will bend wires into hooks to lift a small bucket of food from a vertical pipe. You can even buy puzzles for your dog where it has to learn to uncover a treat by moving plastic sliders.

So it might seem obvious that animals 'know' physics, be they chimpanzees, crows or any of the animals in this book. But the danger when seeing animals act as we do is to assume that they must think the same way. If we see a big box, we 'know' it'll be harder to lift than a small box, but are we right to assume a chimpanzee thinks that too? Based on meticulous observations of chimps wielding tools, Povinelli's conclusion is – as Gershwin once put it – 'it ain't necessarily so'. Chimpanzees aren't even as good as young children at developing an understanding of the physical world.

What's more, chimpanzees (and most of us, for that matter) don't understand physics in all its guises, hence the term 'folk physics'. We build up a picture of the world that works when our lives are on the line, even if that picture's not always true. People typically get the answer wrong when asked how an object will fall if they drop it as they run. Most think the object falls straight to the ground, though in reality it falls forward in a parabolic curve. Povinelli's point is that we so rarely drop objects when running that it doesn't matter if we get it wrong. In most situations, folk physics is fine, but we shouldn't fool ourselves into thinking we 'know' physics.

David Beckham was a master at bending a football into the goal round a wall of defenders, but he's never going to win a Nobel prize for his insight into how turbulence alters the spin of a sphere. Like the bee and the seahorse, Beckham could get turbulence to do exactly what he wanted without knowing – physics-wise – what he was doing. From flying with small wings to creeping up on your dinner to scoring a hat-trick, the key is to get physics to work for you, because if you can, you stand more chance of winning. And in life, it's only the winners that survive.

Acknowledgements

We couldn't have written *Furry Logic* without the help of the following people, to whom we're both hugely grateful.

To all the scientists who've been so generous with their time and knowledge, as well as being lovely people to chat to or correspond with, in approximate order of first appearance: Rick Shine, Andrew Dickerson, Claudio Lazzari, Masato Ono, Masami Sasaki, Aaron Rundus, Helmut Schmitz, Stephen Wroe, David Kisailus, Sheila Patek, Alyssa Stark, John Bush, Mark Denny, Roman Stocker, Michael Nauenberg, Sunny Jung, Brad Gemmell, Charlie Ellington, Matt Wilkinson, Angela Freeman, Holger Goerlitz, Bruce Young, Leo van Hemmen, Caitlin O'Connell, William Turkel, Kenneth Catania, Daniel Robert, Ken Lohmann, Marian Plotkin, Rüdiger Wehner, Matthias Wittlinger, Martin Stevens, Keita Tanaka, Shelby Temple and Sönke Johnsen. Thank you for sharing your expertise, and apologies if we've inadvertently introduced any errors.

To our team of test readers: Mike Follows, Holger Goerlitz, Tania Hershman, Patrick Kalaugher, Maxim Kosek, Bernd Kramer, David Pye, Vijay Shah, Su Smith, Marric Stephens, Shelby Temple and Kate Watt. Thank you for your patience and intelligent insights. To Lotte Kammenga for introducing Liz to the explanation of flight in *Cabin Pressure*.

To everyone at Bloomsbury, including Jim Martin and Anna MacDiarmid for commissioning and editing this work, Aaron Gregory for the fantastic illustrations, and Marc Dando for the expert line diagrams.

Finally, a special note of thanks to *Physics World* features editor, Louise Mayor, who proposed and put together a special issue of the magazine on 'animal physics' in 2012. It was the success of that issue that made us realise what a wonderful idea it would be for a book.

Index